TONY GUNTON

THE PENGUIN DICTIONARY OF INFORMATION TECHNOLOGY

PENGUIN BOOKS

PENGUIN BOOKS

Published by the Penguin Group
Penguin Books Ltd, 27 Wrights Lane, London W8 5TZ, England
Penguin Books USA Inc., 375 Hudson Street, New York, New York 10014, USA
Penguin Books Australia Ltd, Ringwood, Victoria, Australia
Penguin Books Canada Ltd, 10 Alcorn Avenue, Toronto, Ontario, Canada M4V 3B2
Penguin Books (NZ) Ltd, 182–190 Wairau Road, Auckland 10, New Zealand

Penguin Books Ltd, Registered Offices: Harmondsworth, Middlesex, England

First published by NCC Blackwell Ltd 1990
Second edition 1993
This edition published in Penguin Books 1994
3 5 7 9 10 8 6 4

Copyright © Tony Gunton, 1990, 1993
All rights reserved

The moral right of the author has been asserted

Printed in England by Clays Ltd, St Ives plc

Preface to first edition

The problem to be faced when producing a dictionary about information technology is that the topic is so difficult to pin down. In the first place, it is one of the fastest-moving technologies we have ever known, driven along at relentless speed by advances in microelectronics that seem to have no limit. As an inevitable result, new terms are invented and old ones become redundant, almost weekly.

It is also a very versatile technology. It started under the much more humble label of *data processing*, but for the last decade has been reaching out its tentacles, both into neighbouring long-established industries such as office equipment and telecommunications, and into more distant territory such as broadcasting (for example, *teletext, data broadcasting*); printing and publishing (*desktop publishing*); manufacturing (*computer-integrated manufacturing, flexible manufacturing systems*).

To cope with the pace of change in terminology, I have tried to give the emerging technologies which seem to me to have most promise more space than their present status might seem to justify. To cope with the spread of information technology, I have had to decide where it finishes and where the technology whose territory it is invading begins. So it will be helpful if I explain at the outset what the dictionary covers, beginning by saying who it is aimed at.

Who is it for?

Information technology used to be the preserve of the specialist, but now it is an everyday experience for many people, whether through the banking machine they use to get cash, or through the personal computer they have on their office desk or at home.

This dictionary is intended for laypeople such as these as well as for computer specialists, including both complete newcomers and those already involved with information technology as non-specialists, whether at work, as a leisure interest, or in pursuit of training or education. Throughout, I have tried to make the definitions as clear as possible for

iii

such a reader. Generally, this has not meant abandoning technical precision, but on occasions where I felt greater technical precision would obscure the meaning, clarity has come first.

But, as mentioned above, information technology covers a very wide
field, and one that is still expanding and changing rapidly. The dictionary
will also be valuable for computer specialists who wish to keep abreast of
the many developments within their own industry.

I have tried to make each entry stand on its own as far as reasonably
practical, rather than requiring readers to wade through a number of other
references to disentangle the sense of the particular word they want. But I
have also provided cross-references for those who do want to place words
in their full context. Some key terms do recur again and again in definitions though, and unfortunately several of these are widely used with
differing meanings. Readers may therefore find it helpful to look at the
short explanation of how I use some of these terms, given below.

What the dictionary covers

But first let me outline what territory the dictionary covers:

* The base technologies of information technology, starting at the level
  of the integrated circuit, popularly known as the silicon chip, and
  moving up from there to the computers, storage devices, terminals,
  and telecommunications networks in which they are embodied. All of
  those technologies have been around for decades, but all are undergoing change. Notably, storage devices based on optical technologies
  are arriving to challenge the magnetic disk (see *optical disk* and
  document image processing), and the methods used to interface people
  with computer equipment are at last starting to catch up with the needs
  (see *graphical user interface*). Meantime, telecommunications is
  undergoing massive change as new technologies arrive to cope with
  the desktop processing revolution (see *local area network*).
* The way the information technology industry operates. This includes
  the different types of organization supplying products and services,
  and the terms used to describe their marketing methods and tactics (for
  example, *bundle*, *upwards-compatible*). It also includes the various
  specialists in the *information systems department*, and the activities
  they undertake.
* Applications within the 'traditional' areas of data processing – what
  they do, how they operate, how they are described. This is at the level
  of key concepts and issues arising from the application of computers
  that may affect anyone working in business, non-specialist or specialist, and especially covering the key technologies of *database management* and *systems development*.

- What I call 'end-user systems' – areas such as personal computing and office automation where the end-user has much of the discretion in deciding how to use the equipment. Here I cover the terms used to describe the features and facilities of the applications on offer, applications such as *spreadsheet*, *word processing*, and *electronic mail*.
- Programming is the key to the versatility of information technology and is now becoming steadily less and less of a black art, through languages like *BASIC* on personal computers, and the applications packages into which simple mini-programs can easily be embedded. So I cover some key programming terms, describing concepts and methods at a level that will help the interested outsider, the occasional programmer, or anyone who has to deal with programmers.
- Among the emerging technologies, I have already mentioned *optical disk* and *document image processing*. To those I would add in particular *expert systems*, *hypertext*, and *object-oriented programming*.
- I have also picked out some key trade names that already have achieved, or are likely to achieve, the status of *de facto* standards, and the more important among a plethora of abbreviations and acronyms.

A note on basic terms

It is the purpose of a dictionary to explain how terms are used, rather than to lay down hard-and-fast definitions of how words *should* be used. Unfortunately, however, many key terms are used loosely, whether by carelessness, or deliberately by those more interested in creating a marketing impression than in fair use of language. This being so, it is important that I should explain how I have used some key terms that recur frequently within the definitions.

(1) Systems within systems

First, there is the question of how to describe systems, which form a hierarchy one within another (see figure below). As the figure illustrates, I have used *information system* to mean a complete information system in the business sense, including the people that use information and the procedures they adopt. Information systems use long-established technologies such as paper filing systems as well as information technology, and serve to process and distribute the information that organisations use to plan, monitor and control their activites. Within information systems, we find *computer-based systems* (which might equally well be called *IT-based systems*), consisting of a computer system, and the people and procedures that make use of it. A *computer system*, then, is a combination of devices, centred round a computer, that are interconnected and operate in cooperation with one another – the computer along with its peripherals and terminals, the software it runs, and the data it processes. The *computer* itself, finally, comprises the processor, memory and input/output logic, plus a device used to control it, such as a visual display screen and keyboard on a small computer or a console typewriter on a larger one.

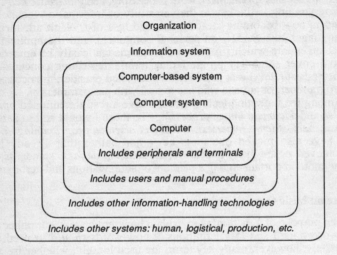

(2) Data, information and knowledge

Clearly the term *information* is central to any understanding of information technology. Do we really know what it is? If we do know, we certainly do not agree about the correct definition. Some place information in a hierarchy between raw facts, referred to as data, and knowledge. Data is transformed into information when it is put into the correct context and related to a particular problem or decision. On this basis it can be defined as 'facts to which a meaning can be attached'. However, some distinguished academics argue that information may be considered a physical entity in its own right, just like matter and energy, although it clearly is not as 'real' as they are.[1] They point out that all organized structures contain and may transmit information – DNA, for example, from the organic world, or a silicon chip from the inorganic world.

The term is also used in a more general sense to encompass all the different ways of representing facts and events within information systems, appearing in various forms including data (structured facts), text

---

1   See Tom Stonier 'Towards a new theory of information', *Telecommunications Policy*, **10**, 4, Dec 1986.

(readable words), speech, image and video (still and moving pictures). I have chosen to use the term in this sense, partly to avoid controversy, partly because we simply do not have another word serving this purpose. Following on from this definition, data is just one form of the information that computer-based systems handle, consisting of highly structured numbers and text.

Is information processing the end of the story? With advances in artificial intelligence research, we are now told that *knowledge processing* is possible also. In its established sense, knowledge is defined as familiarity or understanding gained through human experience, from which it would seem to follow that computers can neither possess nor process knowledge. What they certainly can do is process information so organized that they are able to mimic some of the processes of human thought, representing some of the subtleties of deduction and inference. See, for example, *knowledge base* and *knowledge elicitation.*

Spellings and sequence

I have tried to use the spellings that are in widest use, and that means adopting a number that are of US origin such as *disk, program* and *analog*.

The entries have been arranged according to strict alphabetical sequence of each word and, for the many terms composed of two or more words, the sequence is determined by the first word, then by the second, and so on. Here I have treated hyphenated words as if they were two separate words. This means, for example, that *bit-mapped display* comes after *bit map* and *bit map scanning*. A few terms begin with a numeral, and here I have given the numeral the signifance of the corresponding word (e.g. *3270* and *4GL* are positioned as if they began with *thirty* and *four* respectively).

Tony Gunton
Upminster
December 1989

Preface to second edition

It is just over three years since the first edition of this dictionary was compiled and in that time I have added several hundred new terms, revised about as many of the original definitions, and discarded just a few entries. Some of those changes, I must admit, were omissions from the first edition plus a few corrections, but most of them reflect the continuing advance of information technology into new territory.

As the technology advances so, naturally enough, new terms are invented to describe what it does. Shortly afterwards, new standards emerge for the formatting and transfer of information from machine to machine and from site to site. As the population of computers continues to grow, so this aspect of innovation becomes more and more important since people want to go on exploiting many of their current applications while moving on to take advantage of new possibilities. As many of the new terms relate to this requirement for (in the jargon) *interoperability* as they do to the innovations that have thrown a new spanner into the *compatibility* works.

If new terms are anything to go by, the greatest advances have been in the field of *multimedia*. Today's top-of-the-range personal computers can drive large photograph-quality colour display screens; process and store scanned images of similar quality; and capture and play back clips of speech or video. Multimedia was being talked about in 1989; in 1993 it is available to anyone who can think up something useful to do with it.

Object orientation also was being talked about in 1989; in 1993 no software or systems salesmen worth his or her salt can afford to leave the term off the brochure or out of the sales presentation.

Telecommunications moves forward too. A great deal has been done to help all those personal computers on office desks to take advantage of the huge transmission capacity available in the digital networks hastily being installed all round the world

Tony Gunton
Upminster
April 1993

A

A/D Abbreviation of *analog-to-digital*.

ABI Abbreviation of *application binary interface*.

abort To stop a program, or some other computing activity, before it reaches its intended conclusion.

absolute address A number that identifies the address in hardware terms of a location in a computer's memory. Memory locations are normally numbered from zero upwards, either in bytes or in words.
    Compare *relative address*.

abstract data type A data type consisting of a structure for the data item, plus the operations associated with it. Thus in a manufacturing application, the abstract data type 'part' might have the operations 'release', 'order', 'stock', 'sell' defined for it. Contrast with a conventional data type, which defines only the structure of the data.

accelerator card An expansion card containing a processor that shares the work normally performed by a personal computer's main processor, thus improving processing performance.

acceptance test A test organised by the intended users of a computer-based system before it goes into service, designed to determine whether or not it meets their (previously specified) requirements.

access control Mechanisms included within computer-based systems, and associated procedures, designed to ensure that individuals only gain access to those facilities that they are authorised to use.
    See also *password protection*.

1

access method In general, the way in which records on a mass storage file are retrieved and updated. Also used to describe the software components that enable applications programs to do this. In this case, access method is usually preceded by a description of the file organisation, or in other words the method used to locate records, as in *indexed sequential access method*.

access network The part of a communications network that enables users to connect to it and request its services (as opposed to the *transport network* that serves to carry the information).

access path The means of tracing a computer system resource (a data file, for example) with a multilevel address, such as where the first part of the address identifies a disk, the second part a sub-directory or folder on that disk, and the third a particular entry in the sub-directory.

access time The time needed to retrieve data from a storage device, measured from when the command to read from it is issued until the data is transferred into memory ready for processing. Access time has three main components: (1) seek time, which is the time the storage device needs to position the read/write heads; (2) latency, in other words rotational delay; and (3) read time, which is the time taken to transfer data from the storage device into computer memory. Additional delays may be introduced if the operating software has to wait for a channel to become available to exchange commands and data with the storage device.

accounting routines Routines within an operating system that measure and record the use of a computer system's resources. The information they produce may be used to plan upgrades to the configuration or to work out charges for users.

accumulator A special register used for arithmetic calculations, forming part of the arithmetic/logic unit of a processor.

ACK Abbreviation of *acknowledge character*.

acknowledge character (ACK) A control character used in data communications to indicate that a message has been received correctly. On receiving this character in reply to a message sent down a transmission line, the sending device knows that it can safely continue with further processing, such as sending another message.
Compare *negative acknowledge*.

ACM Abbreviation of *Association of Computing Machinery*.

acoustic coupler A low cost device used to connect (especially portable) personal computers and terminals to the public telephone network, so that data can be transmitted.

The telephone handset is placed in sockets on top of the device. It converts digital data received from the computer or terminal into audible sounds which are sent via the handset, and vice versa.

Compare *modem*.

action diagram A way of representing the processing logic of a program. The basic construct of action diagrams is a square bracket, of which there are various types to represent different control structures and which indicate the possible flow of control within the program. Each bracket surrounds a group of actions, which are described by English-like statements.

active Used of systems that take the initiative to advise their human users of events. An active electronic mail system, for example, is one that alerts a subscriber that a message has arrived, rather than putting it in a electronic mailbox and waiting for the subscriber to collect.

Compare *passive*.

active matrix A technology for liquid crystal display (LCD) screens, used on some portable computers. It is a relatively expensive technology but brighter and with a wider viewing angle than cheaper alternatives.

Compare *supertwist*.

active star A star topology for a network in which the branches of the star are connected via an active component – a controller – that forms the hub of the star. When it receives a signal from one of the branches, the controller forwards it only along the branch to which it is addressed, boosting the signal as it does so. This means that longer wires and more devices can be supported on an active star network than on an equivalent passive star.

activity decomposition diagram A diagram used during the design phase of a computer application. It represents an application in terms of the activities that are performed, each level in the diagram showing these at an increasing level of detail.

activity rate The proportion of records in a data file accessed in a given period.

Compare *hit rate*.

Ada A programming language developed for the US Department of Defense. It is designed for real-time systems and particularly for command and control systems such as are used by the military. It is named after Ada, Lady Lovelace, who first formulated the principles of programming.

adaptive allocation See *dynamic allocation*.

adaptive interface An interface that changes depending on the level of skill of the user,

3

for example by using succinct rather than lengthy explanations or prompts.

add-in card See *expansion card*.

address (1) A number or name identifying a particular computer resource, such as the whereabouts of an item of data in memory or of a record on a direct access storage device such as a disk. See addressing.

(2) Part of a program instruction that specifies where in memory an operand is to be found. The other parts of an instruction are the operation code and the operand(s).

(3) An identifiable device or physical location in a network of devices connected by transmission links.

address buffer A special memory location where a processor stores the address of the next program instruction to be executed. Also known as control register and program counter.

address field A data field, occupying a fixed position within a message or frame, that contains the address of the receiving device. An address field is needed whenever a number of devices share a communications line (as on multipoint lines) or a network (such as a packet-switching network).

addressing The method used to identify individual items within any computer system resource, such as physical records on a direct access storage device or devices attached to a communications network. Addressing schemes may be single level as for memory, where words or bytes are numbered sequentially from zero upwards; or multilevel, where the first part of the address identifies a path, the second part a branch off that path, and so on. Applications programs rarely use physical addresses, where the address corresponds to the hardware address of the device or record, but most often use logical addresses, which the operating software translates into physical addresses. Often, the logical address is exchanged between the applications program and the operating software when the resource is assigned, such as when a file is opened or when a call is established to a remote device across a network. This logical address is then used in all messages or commands until the resource is released again. The advantage of logical addressing is that applications programs do not need to know precise details of the resources available at the time they are running.

ADPCM A widely-used format for encoding audio signals for processing and storage by computer. It is embodied in two leading formats for compact disk, CD-I and CD-ROM XA.

advanced manufacturing technology (AMT) An umbrella term covering all modern computer-based production technologies, including computer numerically controlled (CNC) machine tools, flexible manufacturing systems (FMS) and computer-integrated manufacturing (CIM).

agent A system component (usually software) that takes responsibility for completing one stage of a chain of tasks. For example, CCITT's X.400 standard for electronic mail defines a 'message transfer agent'. An electronic mail message may cross a number of national networks before it reaches its destination, and in each of those a message transfer agent will accept the message either from the originator or from the preceding agent in the chain, and then take responsibility for delivering it to the next agent or to the final recipient.

AI Abbreviation of *artificial intelligence*.

AIFF Abbreviation of *audio interchange file format*.

Algol Abbreviation of algorithmic language, a high-level language used mainly for scientific applications. Algol, the first of the block-structured languages, was introduced in 1958.

allocate Place under the control of a program. System resources such as an area of memory or a peripheral unit may be allocated to a program by the operating system as a result either of a request from the program or a command entered by a computer operator.

alphageometric A method of displaying alphanumeric characters and graphic shapes, by generating them from geometric instructions (known as picture description instructions) transmitted to the visual display device. This method permits the device to display line drawings, colour-filled polygons and approximately curved shapes, as well as text. It is used in videotex systems.

alphameric See *alphanumeric*.

alphamosaic A method of displaying alphanumeric characters and graphic shapes, by generating them from a limited number of mosaic elemental shapes. This permits the terminal to display crude shapes, as well as text, but curved or diagonal lines take on a staircase effect. It is used in videotex systems and for broadcast teletext such as the BBC's Ceefax and ITV's Oracle.

alphanumeric Consisting of alphabetic and numeric characters. Sometimes used to include additionally the special symbols (punctuation, etc.) such as are found on a typewriter keyboard.

alphaphotographic A method of displaying text and picture-quality graphics, assembled from picture elements transmitted individually to the display terminal. This method is used in some videotex systems, and makes heavy demands on the transmission link and the decoder.

alternate path routing Selection of a path through a communications network according to prevailing circumstances, rather than in a fixed, predefined way. Alternate path routing is used in switched networks, such as packet-switching networks, to optimise the throughput of the network and to ensure that failures of network components have a minimum impact on service to network users. When deciding which route to use for a particular message, the switching nodes will take account of known equipment faults and of the loading of the alternative paths that are available.

Alvey programme A programme of research into advanced information technologies, launched by the UK Government in May 1983 with joint state and industry funding. John Alvey was chairman of the committee whose report, published in October 1982, recommended to Government that such a programme should be launched. It identified four technical areas in which major advances were required: (1) software engineering; (2) man/machine interface; (3) intelligent knowledge-based systems; and (4) very large scale integration (VLSI).

The first phase of the Alvey Programme finished in 1987, when it was restructured by Government and, effectively, downgraded.

American National Standards Institute (ANSI) The US national standards-making body. It has been particularly effective in establishing standards for programming languages and database management systems.

AMIS Abbreviation of *audio media integration standard*.

AMT Abbreviation of *advanced manufacturing technology*.

ANA Abbreviation of *Article Numbering Association*.

analog In the form of continuously variable physical quantities. Contrast with digital, where information is expressed as a series of discrete numeric values. Digital computers, by definition, can only handle information in digital form, which means that analog signals must be converted to digital form before they can be processed.

Today's public telephone networks, by contrast, are designed for analog traffic – waveforms representing sound – although they are now gradually being converted to digital working, This is why modems are needed for data transmission – they convert digital data into analog signals for transmission, then reverse the process on receipt. Analog working has a disadvantage for normal telephone traffic as well as for computer data. Because the signal varies continuously and unpredictably, it is difficult to reconstruct it when it is distorted by interference. With digital signals, on the other hand, consisting of a series of 0s or 1s, it is much easier to reconstruct the signal exactly as it was sent.

analog computer Machines designed to perform calculations based on physical quantities rather than, as is the case with digital computers, based on coded text or numbers. For example, in an electrical application the input variables might be voltages; in a mechanical application, the angular rotation of shafts or gear wheels. These inputs are continuous and vary over time, and an analog computer processes them immediately they occur to produce output in the form of a graph or a trace on a cathode ray tube, or to control directly the operation of some other machine or process. Analog computers are used in science and industry to simulate or control physical systems and processes. They are also used for research into design problems, particularly where a solution is required speedily. Unlike digital computers, they cannot store large quantities of data. They are programmed by using a plugboard to interconnect various hardware devices, and this makes them much less flexible than digital computers which hold programs (software) in memory.

analog-to-digital conversion (A/D) The conversion of a continuously varying signal into a series of numeric values. Analog-to-digital conversion is used, for example, to convert the waveform of speech or music signals into a digital representation for recording on media such as compact disks.

analog-to-digital converter See *digitiser*.

analogue See *analog*.

analyst/designer workbench A computer-based tool that helps systems analysts or designers to design computer applications and programs, and to produce the documentation that describes them.

analyst/programmer See *programmer/analyst*.

AND circuit A circuit with two or more inputs and one output whose output is high if and only if all the inputs are high.
　　See also *AND operation*; *logic element*.

AND operation A program instruction or statement that performs a boolean AND operation on two data fields. As shown below, a one bit is set in the output field where the bits in both of the input fields are one, otherwise the bit is set to zero.

Input		Output
0	0	0
0	1	0
1	0	0
1	1	1

anisochronous transmission See *asynchronous transmission*.

ANSI Abbreviation of *American National Standards Institute*.

antecedent-driven reasoning See *forward chaining*.

antecedent rule The type of rule that is applied in a forward-chaining expert system. It takes the general form 'IF A applies, THEN draw conclusion B'.
Compare *consequent rule*.

antialiasing A technique used by painting programs to improve the appearance of artwork. It consists in blending the edges of a line or a character of type into the background shade or colour, making the transition appear softer and more natural.

antiglare filter A supplementary screen fitted over a visual display screen, designed to reduce reflection and glare from the screen.

API Abbreviation of *application programming interface*.

APL Abbreviation of A Programming Language, a high-level language specially suited to handling multidimensional arrays. It provides a set of powerful operators and also allows programmers to define their own new operators.

APP Abbreviation of *application portability profile*.

application A particular problem to which information technology is applied. Also used to refer to the hardware and software which is used to express the application, as in 'they demonstrated the stock control application'.

application architecture A high level model showing the applications that do or will serve an organisation, and how these relate to one another in terms of the data flows between them. Such an architecture is normally produced as a result of an information systems planning study and would subsequently be used as a blueprint for developing a suite of integrated applications.
Compare *conceptual data model*.

application binary interface (ABI) A definition of the format of object (i.e. compiled, also referred to as binary) programs intended to run on a particular type of computer. Contrast with *application programming interface*, which defines the format of the source program.

application generator A program, usually in the form of an applications package, that is used to create applications programs. The user does this by entering various details about the application to be generated, such as the formats of display screens and data files. Compared with a conventional programming language, application generators are easier to use but both less versatile and less efficient in their use of hardware resources. One of the most widely-used application generators is Ashton-Tate's *DBase*, available on many personal computers.

Compare *fourth generation language*.

application layer The seventh and highest protocol layer of the OSI reference model, defined by ISO. The application layer is the level at which, from the user's point of view, useful work is done, all the lower levels being there to administer the preliminaries and the communications activities. ISO's work and the reference model are aimed at standardising the protocols that are used. Much of the processing at this level will be specific to the application concerned and therefore not amenable to standards, but some activities do recur and can be standardised, such as transferring files or sending electronic messages. It is also hoped to standardise network management functions, such as directory and security services.

See also *data link layer*; *network layer*; *physical layer*; *presentation layer*; *session layer*; *transport layer*.

application level See *application layer*.

application literate See *computer literate*.

application package A program, or a set of programs, designed to meet the requirements of a particular class of application. The application might be defined in terms of the industry to which a user organisation belongs and/or the aspect of its operations that the package addresses, as in the case of a travel agent administration package or a payroll package, respectively. Alternatively, it might be defined in terms of the type of task that the package supports, as in the case of a word processing or a spreadsheet package. Much of the operation of an application package will be fixed, reflecting the requirements it is designed to meet, but normally particular users can modify the way it operates, or the format and frequency of its input or output. This is done by selecting different options from a range of choices offered by the package, either when the package is first installed or when it is run.

application portability profile (APP) A set of standards formulated by the US National Institute of Standards and Technology (formerly the National Bureau of Standards), intended to be applied throughout the US federal sector. The standards are aimed at enabling applications programs to run on a range of different computer equipment.

See also *functional profile*; *open systems interconnection*.

application program A program designed to meet the requirements of a particular individual or organisation, and normally developed specifically for that individual or organisation. This can be contrasted with application packages, designed to meet the requirements of a class of individuals or organisations; and with operating software, which helps any user of a particular type of computer equipment to manage its operation and exploit its capabilities.

application programming interface (API) A definition of the format of a source program acceptable for a particular range of computers. Contrast with *application binary interface* (ABI), which defines the format of the object program.

application specific integrated circuit (ASIC) An integrated circuit fabricated for a particular application, such as for a digital watch or to control a household appliance. Can be contrasted with general-purpose integrated circuits such as microprocessor chips or memory chips.

applications backlog The new applications, or amendments to existing applications, that are waiting to be implemented by an organisation's information systems department. In many large organisations, there is an applications backlog several years' long.

See also *hidden backlog*.

architecture The way in which design, hardware and software interact in order to provide a planned level of capability and performance. The term is applied both to individual computers, and also to an information system that may consist of a number of interlinked computer systems and terminals.

Manufacturers design individual models of computer to meet the needs of a particular segment of the market. The architecture of a particular model is the way in which the hardware and operating software co-operate to meet design objectives, such as processing throughput for the anticipated workload, reliability, ease of maintenance, ease of programming and operation, manufacturing cost, and so on.

At the level of an information system as a whole, the architecture defines how the components of that information system will interact with one another in order to exchange information and to run programs in a co-ordinated manner. The market leader, IBM, sanctified the term in 1981 by announcing systems network architecture (SNA). SNA defined an overall structure for networks of computers and terminals and a set of rules (known as protocols) governing the mechanisms they use to exchange information.

In the public arena, the International Standards Organisation (ISO) has defined what it called a 'reference model for open systems interconnection' (often referred to as 'the OSI reference model'). This could equally well have been called an architecture, and has since been used as the basis for a range of international standards for information technology products, intended to enable products from different suppliers to work together.

archival storage Storage media used to hold information for archiving purposes, or in other words for security reasons or for occasional access rather than for regular processing or retrieval. Archival storage normally takes longer to access than other forms of mass storage, but the storage cost is lower. Magnetic tape and microfilm devices have been used for this purpose in the past, but are now being superceded by WORM (Write Once Read Many times) optical storage devices.

archiving Long term storage of information on electronic media. Information is archived for legal, security or historical reasons, rather than for regular processing or retrieval.

area A logical subdivision of a database, consisting of one or more files.

areal density The quantity of information that can be stored in a given space on a storage device such as a magnetic disk, typically expressed in megabits (millions of bits) per square inch.

argument (1) A variable included in a call to a function, and which is used to determine the value of the function.
   (2) The fixed point part of a floating point number.

arithmetic operator Defines an operation involving arithmetic instructions – add, subtract, multiply, divide, etc. Both the operands on which the operation is performed and also the result will be numeric. Compare *logical operator*.

arithmetic overflow Where the result of an arithmetic calculation performed in a program is too large for the register(s) or memory location(s) intended to hold it. When this happens, the processor either sets an overflow indicator or generates a message to warn the program concerned. It can then decide what remedial action to take.

arithmetic shift An operation which causes the bits in one or more words to shift a specified number of bits in either direction. The sign bit remains as it was before the shift. In other words, this is a means of multiplying or dividing a binary number by a power of two. Compare *logical shift*.

arithmetic/logic unit The part of a central processor that decodes program instructions and carries out the required processing – arithmetic calculations, movement of data to and from memory, and so on.

ARPAnet A resource-sharing network established by the Advanced Research Projects Agency of the US Department of Defense in the early 1970s. ARPAnet was the prototype for the packet-switching networks that have since been installed throughout the world.

ARQ Abbreviation of *automatic request for repetition*.

array A data structure that contains elements all of the same type, such as a series of numbers or a series of text strings. Each element in the array can be accessed directly by means of an index value (known as a subscript) specifying the relative position of the element within the array.

array processor A form of parallel processing computer system designed to process large arrays of numbers. It consists of a set of processors that operate simultaneously, each of them working on one element of the array.

Article Numbering Association (ANA) An association of UK business organisations involved in retail product distribution. Members include retailers, wholesalers and their suppliers. The ANA defined the Tradacoms standards for the format of electronic messages used for stock ordering, invoicing and similar business transactions. Subsequently, the ANA was instrumental in setting up Tradanet, an electronic data interchange network for such transactions.

artificial intelligence (AI) The goal of artificial intelligence work in the information technology field is to develop computer systems that exhibit some of the intelligence characteristics of human beings. It is an intellectual hybrid built on the disciplines of philosophy, linguistics, mathematics, electrical engineering and computer science. There are several sub-areas within AI, of which the most important are expert systems, natural language understanding, computer vision, knowledge representation and learning systems. A different approach is taken to artificial intelligence by cognitive scientists where the aim is to study the functioning of the human brain.

ASCII Pronounced 'askey', this is a contraction of American Standard Code for Information Interchange. It is the character set (i.e. the codes identifying different characters) used by almost all personal computers, and by many visual display terminals. It is also used almost universally for access to public information retrieval and messaging services. Originally developed in the US, it has been extended and ratified by international standards-making bodies such as ISO and CCITT.

It is a seven-bit code, giving 128 unique combinations or characters. 32 of those are reserved for control functions relating to printing or transmission, such as backspace, carriage return, end of text; and the remaining 96 give the alphabetic (upper and lower case), numeric and special characters (such as punctuation marks) found on a normal typewriter keyboard. For full details see Appendix 1.

ASCII characters are normally stored as eight-bit (that is, one-byte) characters. For transmission over telecommunications lines, the eighth bit is used as a parity bit, but some computers use the eighth bit internally to provide an extended character set for applications such as word processing and desktop publishing. This makes it possible to represent national characters using accents; special symbols such as ©; greek letters and other symbols used in mathematics; and special typesetting symbols such as en-dash.

ASCII string See *string*.

ASIC Abbreviation of *application specific integrated circuit.*

aspect ratio The proportions of a visual display screen, expressed as the ratio of its width to its height. The usual standard is 4:3, although A4 (about 3:4) screens are sometimes used for word processing and desktop publishing applications.

assembler A program that translates a source program written in an assembly language into an equivalent program in machine language, so that it can be executed by a computer.

assembly language A programming language whose statements correspond more or less one-to-one with the machine language instructions they generate. This means that assembly languages can generally only be used on one particular model or type of computer. Assembly languages help the programmer in two main ways:

(1) They use mnemonic codes to identify instructions (for example, STO for a store instruction, MPY for multiply, etc.), which are easier to remember and understand than the machine language operation codes.

(2) Addresses of data fields or instructions can be identified by symbolic names or labels, which the compiler translates into memory addresses. This means that data fields or statements can be added or removed without the programmer needing to recalculate addresses.

Most assembly languages also have a macro capability, whereby a single statement generates a predefined sequence of statements automatically.

Compare *high-level language.*

assign (1) Of a peripheral unit, place under the control of a program. Similar in meaning to allocate, but perhaps implying more permanence.

(2) In programming, give a value to a variable.

assignment statement A statement in a high-level language that gives (assigns) a particular value to a variable.

Association of Computing Machinery (ACM) An association of people professionally involved with computing, based in the United States. The ACM runs regular conferences on a number of computing topics.

associative network A method of representing knowledge, as a preliminary to recording it in the knowledge base of an expert system. It shows the associations between terms such as, for example, that 'Fred' is a man and that 'kick' involves use of the foot. This enables an expert system to draw wider inferences from the facts with which it is presented. For example, on being told that 'Fred kicked the ball', an expert system with these associations represented in its knowledge base could also answer the question 'Did any man use his foot?'. See also *expert system; knowledge base.*

associative retrieval Where data records on direct access storage such as disk are located by direct reference to their contents, rather than by means of an address. Conventional storage devices use the latter method – either the address of any given record on the storage device is computed from some part of it which identifies the record uniquely (the key field), or alternatively index records containing key fields and associated record addresses are searched. With both these methods, applications programs must either know the key field to retrieve a given record, or must search the whole file checking one record after another. With associative retrieval, by contrast, they can retrieve records based on any part, or combination of parts, of the record contents.

Associative retrieval is sometimes supported by special direct access storage hardware capable of reading many records in parallel, whereas conventional hardware can only read one record at a time. This is commonly referred to as content-addressable memory.

asynchronous terminal A terminal that uses asynchronous transmission techniques to transmit messages.

asynchronous transfer mode (ATM) A standard transport technology underlying high-throughput protocols such as cell relay. It takes incoming traffic (which may be voice, data or video) and divides it into cells for onward transmission, each carrying instructions on where the cell is to be delivered.

asynchronous transmission A method of data transmission in which each character is sent separately, rather than in a continuous, synchronised stream. Each character is framed by start and stop bits, and for this reason it is also known as start-stop transmission. On detecting a start bit, the receiving terminal counts off the succeeding bits of the character at a series of fixed time intervals. Asynchronous transmission is used by slow-speed data entry devices such as teletypes, and also by personal computers. Speeds up to 2400 bit/s, and sometimes higher, can be achieved over the public telephone network. Less commonly referred to as anisochronous transmission.

Compare *synchronous transmission*.

ATM Abbreviation of *asynchronous transfer mode*.

ATM Abbreviation of *automatic teller machine*.

attended operation A mode of operation of a computer system or a terminal that requires a human being to be present.

attenuation The gradual reduction or corruption of a signal transmitted along a transmission line. To overcome the effects of attenuation, repeaters are installed at intervals along a transmission line. These attempt to correct and regenerate the signal.

attribute (1) In database terminology, a characteristic of an entity. Thus the attributes of an entity called 'product' might be product code, product description, minimum order quantity and store location. When represented in a data file, attributes correspond to the data fields in each record.

(2) In a rule-based language used to describe a knowledge base, any term that describes something about its associated context. Thus 'tax' and 'credit rating' might be attributes of the context 'citizen'.

attribute inheritance A database concept whereby entities automatically take on the attributes of entities higher up in the class hierarchy than they are. For example, the database might include a class called 'people', having attributes such as two legs, two arms, two eyes and so on. Any other entities in the database defined as people ('men', 'women', 'children') would implicitly be assumed to have the same attributes unless explicitly defined otherwise, such as where it is specified, for instance, that a particular man had only one eye.

See also *class hierarchy*.

audio conferencing See *teleconferencing*.

audio graphics A form of teleconferencing combining graphics-oriented visuals with voice teleconferencing. Provides a midpoint in terms of cost and power between video and audio teleconferencing.

audio interchange file format (AIFF) A standard for the representation in digital form of audio information such as speech or music.

audio media integration standard (AMIS) A standard protocol for the interconnection of computers and private telephone exchanges (PABXs), for applications such as voice messaging. It is backed by the US Information Industry Association.

See also *computer-integrated telephony*.

audio response See *voice response*.

audiotex A term used to describe applications where a touchtone telephone is used in conjunction with an automated voice-response system. A number of such systems are operated by banks to provide a home banking service. Customers dial in to the service, then, prompted by the voice-response system, use the telephone keypad or a separate keypad to enter details of account number, PIN number, what service they require and so on. Contrast with videotex, where the telephone is used in conjunction with a computer display or a television set.

audit trail A means of reconstructing the sequence of processing carried out by a computer-based system, such as a journal of transactions that can be used subsequently to

15

check whether processing was correct.

authorisation level The status accorded to a user of a computer system, determining which of the services or information available on the system he or she may access.

See also *password protection*.

autoanswer Used of modems which can detect an incoming data call and 'answer the phone' automatically, without human intervention.

autodial Used of modems that can dial a specified number (sent to it by the computer system to which it is attached), then switch into data mode automatically as soon as the called device answers.

automatic hyphenation A feature of a word processing package that hyphenates words automatically at the end of lines. This is either done as text is typed or when the user requests it. Most automatic hyphenation is based on special dictionaries of hyphenated words, stored on disk.

automatic request for repetition (ARQ) A feature of some modems, usually abbreviated to ARQ. These modems automatically attach an error-detecting code to each transmission. The receiving modem checks this code and, on detecting a transmission error, automatically sends a request for the transmission to be repeated.

automatic teller machine (ATM) A device used to dispense cash and for other banking functions, popularly known as a 'hole-in-the-wall' banking machine. Most ATMs read plastic cards which identify the customer and perhaps carry details of credit limits. They are connected by data communication links to the computer systems of banks and other financial service companies, so that credit limits can be checked; stolen cards can be recognised; and transaction details can be recorded automatically.

autosave Automatic saving of the results of processing in a semi-permanent form, such as on disk. Some applications packages that hold a great deal of information in memory during processing, such as word processors, use autosave periodically to minimise the cost to the user of a failure, such as loss of power, that results in loss of the current contents of computer memory.

availability A measure of the time that a computer system is able to provide a service. It is the percentage of time that it is available for normal use, within the total time that it is switched on. Time may be lost as a result of system failure, and for maintenance and repair work. This lost time is known as *down time*.

See also *reliability availability serviceability*.

B

BABT Abbreviation of *British Approvals Board for Telecommunications.*

BABT-approved A designation for equipment that is approved for connection to public telecommunications services in the UK. BABT stands for the British Approvals Board for Telecommunications.

Bachman diagram A diagram used to represent entities in a database and the relationships between them, based on the network model of database structure. Charles Bachman was one of the originators of the database concept, and particularly of the CODASYL proposals that formed the basis for the earliest products.

back-end processor A processor which forms part of a larger computer system and whose function is to run programs at the dictates of the processor controlling the system as a whole. Normally, a back-end processor will be installed in the same room as the controlling processor and will perform a specialised set of tasks, such as for example managing a database.
    Compare *front-end processor.*

back records The records already existing in an organisation before a computer-based system is introduced, and that need to be captured in electronic form so that the new system can operate.

back up As a verb, to create a duplicate copy – of a program or a data file or a complete disk – normally on a different disk or on a different medium such as cassette tape. As a noun (often spelt as one word), the duplicate copy itself. Backups are made as a protection against loss of or damage to the original.

backbone network The part of a network that connects sites and switching systems to one another, rather than connecting the user devices such as computers and terminals.

The latter is termed the site or access network. Also known as trunk or transport network, and very similar in meaning to wide area network.

background printing A feature of some personal computer operating systems that allows you to print a document while continuing to use the computer keyboard and screen for other work.

See also *background processing*.

background processing Low priority processing within a computer system over which the user has no direct control. To control a background operation, the user must bring it to the foreground such as by issuing commands to the operating system.

Foreground/background processing is a simple method, adopted on a number of personal computers, of overlapping lengthy processes requiring little processing time and no user intervention with normal user interaction with the machine. For example, a print run may be started and will run in the background until its completion, while the user continues keying text in the foreground. At any one time there may be several applications running in the background but only one in the foreground.

backing storage See *mass storage*.

backlog See *applications backlog*.

backplane See *motherboard*.

backspace character The control character generated as a result of pressing the backspace key on the keyboard of a terminal or personal computer. The character is sent to the processor controlling the output device to which the keyboard is attached, such as a visual display unit or a printer, which moves the cursor or the print mechanism back one character position.

backup See *back up*.

backward chaining A control strategy used by the reasoning mechanism of an expert system, the inference engine. The program starts by assuming a conclusion or goal, then works backward trying to satisfy all the conditions leading to that goal.

Compare *forward chaining*.

backward reasoning See *backward chaining*.

badge reader A device used in security access and time recording systems to identify people. Badge readers generally read and check a plastic card or some other small identifying object. They may wait for the user to key in an identifying code that must match with the code recorded on the badge, and/or check an expiry date, and/or capture details of the code and the current time for processing or checking later.

bandwidth The frequency range of a transmission channel, usually expressed in kiloHertz (KHz) or megaHertz (MHz). For digital traffic, this defines its carrying capacity. Bandwidth is often treated as synonymous with *bit rate*, defined in bits per second.

bandwidth broker A supplier of telecommunications services who offers end-to-end line capacity between two locations and manages the traffic carried along those lines on behalf of the customer.

bar code An arrangement of lines of varying thickness with spaces of varying length between them, representing a numeric code. Bar codes are used mainly in the distribution industries, to identify products in shops and warehouses. They can more easily be read automatically by light pens or scanners, than can ordinary numbers. In Europe, the European Article Numbering Code defines bar code standards, while in North America the Universal Product Code is used.

bar code scanner A device that reads the bar codes used to identify products in shops and warehouses, and converts them into the numeric codes which they represent.

base address The lowest address occupied by a program when loaded into memory. The processor automatically adds the base address to any addresses generated by program instructions, in order to calculate the actual memory locations (known as absolute addresses) to be accessed. Also used more generally to mean the lowest address of any area in memory, and which must be added to relative addresses within that area to calculate absolute addresses.

base station An installation that enables mobile terminals within a limited range to gain access to a communications network providing broader geographical coverage.

baseband Describes a telecommunications system in which information is directly encoded on to the transmission medium, rather than using a carrier signal. The term is used of modems that can only be used over short distances on private lines, and also of local area networks. Baseband local area networks, usually consisting of coaxial cable, assign their entire capacity for a very brief period of time to a single user device. This makes them suitable for large numbers of low-volume devices, such as terminals or personal computers. They can be contrasted with broadband local area networks, which divide transmission capacity up into a number of channels that are used concurrently.

BASIC Acronym for Beginners All-purpose Symbolic Instruction Code, a high-level programming language supported by most personal and home computers. It was created as an educational tool by Kemeny and Kurtz, maths professors at Dartmouth College,

New Hampshire, in the mid-1960s. Its popularity derives both from its simplicity and because program statements can be entered and checked interactively, at the computer keyboard. Also, BASIC programs are often interpreted (i.e. executed at the same time as they are read), rather than compiled. This makes debugging relatively straightforward, because changes can be made to the program statements as soon as errors are recognised and a test can be initiated immediately simply by typing the word RUN, rather than having to go through a lengthy re-compilation process. Unfortunately, the BASIC language, like COBOL, is another standard language that is not quite standard, since almost every computer manufacturer has adopted a slightly different version, especially for graphics.

See also *true BASIC*.

Sample BASIC program (numbered lines), followed by test run
(Computer output underlined)

```
10 INPUT "Print how many lines"; X
20 FOR I = 1 TO X
30 PRINT I " of " X
40 NEXT I
50 END
RUN
```

Print how many lines? 5

 1 of 5
 2 of 5
 3 of 5
 4 of 5
 5 of 5
OK

---

batch processing Computer processing of information that has been assembled into batches of transactions prior to input. Batch processing was the main method used in the early days of data processing. The batches of transactions would first be punched into cards or keyed directly on to magnetic tape, at which stage control totals would be calculated that could subsequently be checked to ensure that all the transactions in the batch had been processed. These transaction records would then be sorted, ready to be processed against master files held on magnetic tape. Any transactions that could not be processed, for example because they were incomplete or because codes were invalid, would be recycled for correction and reprocessing. Batch processing is efficient in terms of processing power, but has major practical disadvantages. Notably, it introduces delays and complicates the detection and correction of errors in the input data. It is still used for various forms of background processing, such as to generate summary files or to produce

20

reports, but is being superseded by teleprocessing or by distributed processing for the capture and processing of business transactions.

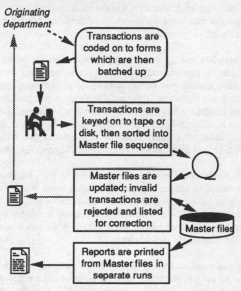

Originating department

Transactions are coded on to forms which are then batched up

Transactions are keyed on to tape or disk, then sorted into Master file sequence

Master files are updated; invalid transactions are rejected and listed for correction

Master files

Reports are printed from Master files in separate runs

batch total A total accumulated by adding together some of the fields in a batch of documents or records. This total is checked at various stages in the processing of the batch to check that all records are present and have been read correctly.

baud A unit of signalling speed of a transmission channel, meaning the number of pulses transmitted. For all practical purposes, baud rate is synonymous with bit rate and is measured in bits (or kilobits, etc.) per second.

BCC Abbreviation of *block check character*.

BCD Abbreviation of *binary coded decimal*.

BCS Abbreviation of *British Computer Society*.

BEGIN statement A statement used in some high-level languages to mark the beginning of the program logic (as opposed to data and other declarations), or the beginning of a block within the logic – see *block-structured language*.

21

beginning of tape marker (BOT) A reflective strip near the beginning of a magnetic tape that marks the point beyond which information can be recorded on the tape. When a tape is mounted on a tape unit, and after it is rewound, the unit automatically winds the tape forward until the read/write heads are positioned beyond this marker, otherwise known as the load point.

benchmark A standard task or set of tasks to be carried out by a computer system, designed to measure its performance with a particular workload. Benchmarks consist either of a specially chosen mix of instructions or of programs specially written to simulate the anticipated workload. They are used both to compare the performance of different suppliers' systems and to help assess equipment requirements. Post Office Work Unit and Gibson's Mix benchmarks were widely used in the past to measure processor performance, but have been superceded by benchmarks that measure transaction throughput, such as Debit/Credit (and derivatives like TP1) and RAMP-C.

Bernoulli drive A disk drive based on the 'Bernoulli effect', which describes how air moving above a surface at a faster rate than air moving below it will cause the surface to lift, like the wing of an aeroplane. Bernoulli drives use a spinning flexible disk. An airflow is applied above and below the disk which keeps it stable at a constant distance from the read/write head placed above it. If the airflow or the power is disrupted, the disk falls away from the heads, thus avoiding damage.

beta test The final stage of testing of a software package, or of a new version of a package, prior to its general release. Beta tests are generally carried out by making the package available to a limited number of trusted customers or trading partners, who use it and report their experience back to the supplier.

bezier curve A type of curve used to describe the shape of characters included in a high-quality font, such as might be used for desktop publishing.

biCMOS An integrated circuit technology that combines bipolar and complementary metal oxide semiconductor (CMOS) technology on the same chip.

binary chop An efficient way of searching for a particular entry in a sequential list. The list is successively divided into equal parts, and after each 'chop' the part containing the entry is selected, by comparing the entry sought with the entry at the point of division. This process continues until a match is found or the list has been reduced to one.

binary coded decimal (BCD) A way of representing numeric characters stored in a computer system. Each digit of the decimal number is stored in four bits, in other words packed two to a byte. Also known as *packed decimal*.

binary digit See *bit*.

22

binary large object A term used to describe database fields containing 'unstructured' (i.e. compared with numbers or text in character form) data such as digitally encoded images or sounds.

binary notation A system for representing numbers in which the base for each digit position is two, rather than ten as in decimal notation. Thus, numbers are represented by a series of digits consisting of 0 or 1. As in the decimal system a displacement of one digit position to the left means the digit is multiplied by a power of ten, so in the binary system displacement to the left means multiply by a power of two. So, for example, '10' represents 2, and '101101' represents $32 + 8 + 4 + 1 = 45$. Binary notation is used within computer systems because their basic components can only distinguish between two states, represented by the presence and absence of an electrical charge respectively.

binary number A number represented in binary form, rather than, say, as a series of characters.

binary portability See *portable*.

binary synchronous communications (BSC) A set of protocols for synchronous transmission of data between computer systems and terminals, defined by IBM. Often referred to as bisync or BSC protocols. Messages consist of a string of 8-bit characters, either in EBCDIC code or in 7-bit ASCII code with one parity bit. Bisync protocols are widely used for transaction processing and for file and job transfer, but are being superceded by more versatile bit-synchronous protocols such as high-level data link control (HDLC), embodied in international standards for packet-switching networks, and synchronous data link control (SDLC), embodied in IBM's proprietary networking scheme, systems network architecture (SNA).

bipolar A type of integrated circuit technology. Bipolar circuits switch more quickly than the other type of technology, complementary metal oxide semiconductor (CMOS), but consume more power and dissipate more heat. Therefore it is more difficult to achieve such high levels of circuit integration. The fastest type of bipolar technology is emitter coupled logic, used in the processor logic of powerful mainframe computer systems.
  See also *biCMOS*.

bisync See *binary synchronous communications*.

bit A contraction of Binary digIT, in other words a digit in binary notation, consisting of either a '0' or a '1' (or Off and On). Binary notation is used within computer systems because their basic components can only distinguish between two states, represented by the presence and absence of an electrical charge respectively, so bits are the basic unit of storage in the memory and on the mass storage devices, such as disks and tapes, used in computer systems. For processing and for display to the user, bits are normally grouped to form binary codes, such as the ASCII codes that represent alphanumeric and other characters or the operation codes used to represent machine language instructions. They are also grouped to represent numbers or memory addresses. Eight bits make up a byte, which is the normal grouping on personal computers.

bit error rate A measure of the quality of a data transmission channel, such as a telecommunications link or a connection between a computer processor and one of its peripherals. Expresssed as a ratio of the average number of bits received incorrectly to the total number of bits transmitted (for example, 1 bit in 10^7).

bit image A collection of bits in the memory of a computer system that represent a rectangular shape. The same shape displayed on a screen is a visible bit image.

This bit image...

```
11111100
11111110     might be displayed
11000011     like this ...
11000011
11111110
11111110
11000011
11000011
11111110
11111100
```

bit map A data field treated as a series of bits, each of which represents an element in some other arrangement of computer resources. A bit map can be used, for example, to control the space on a disk or in computer memory. Each bit represents a sector or some other unit of space, and is set to 0 if the corresponding area is free, or to 1 if it is allocated. Bit maps are also used to represent the contents of display screens (see *bit-mapped display*) and to represent graphic objects such as characters in a font.

bit map scanning See *image scanning*.

bit-mapped display Where each dot (or picture element – *pixel*) of a display screen can be addressed individually, and is represented by a bit (or a group of bits) in computer memory (known as the *screen buffer*). In the case of a monochrome screen one bit is used for each pixel, representing black or white. With a colour or grey-scale screen, a number of bits are used for each pixel, representing the shade or colour. Bit-mapped screens can be contrasted with *80-column display* screens, where the screen is divided horizontally (usually 24 or 25 lines) and vertically (80 columns) into invisible cells, in each of which an alphanumeric character can be displayed. Bit-mapped screens have the advantage of great flexibility, being able to display a combination of text and graphics, but place greater demands on the processor and the software driving them.

bit-oriented protocols See *bit-synchronous*.

bit pattern A group of bits forming a specific pattern with a particular meaning, used for example to represent a character or an instruction.

bit position The relative position of a bit in a group of bits, such as a byte or a word. Bit positions are normally numbered from left to right, beginning from 0, so the rightmost bit position in a byte would be bit 7.

bit rate A measure of the speed of a transmission channel or a modem, expressed in bits (or kilobits, etc.) per second. Also referred to as baud rate. Modem bit rates are sometimes represented by two numbers, such as 75/1200. These represent the speed first of the originating then the receiving transmission channel, which are not always identical.

bit slice microprocessor A microprocessor built up from a number of units chained together by microcode. Each of these units (slices) can operate as an arithmetic/logic unit handling a certain number of bits – for example, 4 or 8 bits. This makes it possible to build a processor of any desired word length.

bit stuffing A technique used in bit-synchronous protocols to enable any sequence of bits to be carried in the information field. All transmissions are in frames preceded and followed by a fixed bit pattern known as a flag, consisting of the bit sequence 01111110. To ensure that that same bit pattern does not occur within the information field of the frame, the transmitting device automatically 'stuffs' an extra 0 bit after any five consecutive 1 bits. These extra bits are removed automatically by the receiving device.

bit-synchronous Describes a type of synchronous transmission protocol in which the information carried may be any required sequence of bits – characters, bytes, binary data. This is in contrast to the *binary synchronous* protocols which prevailed before the arrival of bit-synchronous protocols, where the information had to consist of a string of 8-bit characters. All transmissions in bit-synchronous protocols are in frames, several of which may form a message, depending on length. Each frame is preceded and followed by a fixed bit pattern known as a flag (see *bit stuffing*).

Bit-synchronous protocols such as high-level data link control (HDLC), embodied in international standards for packet-switching networks, and synchronous data link control (SDLC), embodied in IBM's proprietary networking scheme, systems network architecture (SNA), are more versatile than binary synchronous protocols. They can cope with high traffic volumes, such as are found on public packet-switching networks, and are designed for high-speed communication between computer systems, as well as between computers and their terminals.

bit/s Abbreviation of *bits per second*.

bitonal See *monochrome*.

bits per inch (bpi) A measure of the recording density of magnetic storage media, and particularly magnetic tape.

bits per second (bit/s) A measure of the speed of operation of digital equipment that transfers information, such as channels connecting a computer processor to its peripherals or modems connecting devices to transmission lines.

blob See *binary large object*.

block (1) A physical unit of storage for data on mass storage devices. Records (the logical unit of data with which applications programs normally work) are assembled into blocks to make efficient use of mass storage space. On magnetic tape, for example, more space would be used for interblock gaps than for data if records were written to tape physically separate. Blocks on magnetic tape may vary in length up to a maximum defined for a particular file, which in turn defines the length of the buffer into which blocks are read before the records it contains are processed. The length of blocks on disk normally corresponds with the length of the sectors into which the recording surface is divided.
   (2) For block-structured languages such as ALGOL or Pascal, a module of a program containing both a definition of the data fields it uses and the program logic.

block check character (BCC) A character appended to the information field in a transmission block. It contains a value computed from the information field, such as a longitudinal parity check, and is checked on receipt to verify that the information has not been corrupted in transmission.

block diagram A diagram representing a system or a program, in which each logical unit is represented by a rectangle or shape containing descriptive text and the relationship between them is shown by connecting arrows.

block graphics See *alphamosaic*.

block length The maximum length defined for blocks of data in a particular mass storage file, measured in terms of the number of characters, bytes or words of data it contains. Also used to mean the actual length of a particular block.

block-structured language A programming language in which programs are written in sections (called blocks), whose logic and internal variables are only recognised when the block is executed, and are then forgotten again when execution moves on to the next block. This makes it possible to build programs up block by block in a modular fashion, without needing to worry, for example, that the same name has been used for different variables in different parts of the program (see *stepwise refinement*). It also makes it possible to write recursive routines, in other words routines that call themselves. Algol was the first block-structured language, and has since been followed by languages such as PL/1, Pascal and C.

block wiring A term used to describe the organised approach normally used to cable a building for telephone extensions.
Compare *structured wiring*.

blocking factor The average number of records contained in each block of a mass storage file.

blow To establish a program or data on a programmable read-only memory (PROM) chip. Also referred to as *burn*.

board See *printed circuit board*.

body text Text in small sizes, usually between 8-point and 14-point.
Compare *display text*.

boilerplate Standard sections of text, held on direct access storage for automatic incorporation into documents by word processing software.

boolean algebra A system of algebraic notation that expresses logical relationships, just as conventional algebra expresses mathematical relationships. It is named after the mathematician George Boole, who died in 1854. The variables used in Boolean algebra do not stand for numbers but for expressions; the operators which link these variables stand for conjunctions such as 'or', 'and', 'not'; and the results can take only one of two forms 'true' or 'false'. It is this feature of Boolean algebra that makes it a suitable basis for the logic of computer systems, since the electronic circuits from which that logic is built also have only two possible states – either a presence or an absence of a charge, in turn signifying a binary digit that can only be zero or one.

boolean operator See *logical operator*.

boolean variable A variable which may have only two possible values, meaning TRUE or FALSE.

boot The act of initialising a computer system immediately after it is switched on or at the beginning of a session. It usually involves loading part or all of the operating system into memory. It is a contraction of bootstrap, and so called because a small routine is read into memory first or resides there permanently, and 'bootstraps' the operating system in after it.

bootstrap See *boot*.

BOT Abbreviation of *beginning of tape marker*.

bottom-up Beginning with the detail, then progressively assembling the detail into larger units which eventually form the whole. The reverse approach is known as top-down.

bpi Abbreviation of *bits per inch*.

branch A programming construct consisting of a condition that can take one or more values and a list of corresponding options. Depending on the value of the condition, program execution continues with one of these options. A branch may be unconditional (see *GO TO statement*); or it may be a simple choice conditional on the result of a test, such as comparing two numbers (see *IF statement*); or it may consist of a series of options identified by a variable (see *CASE statement*).

breadboard Assemble a prototype of an electronic product by plugging the components into a special board that enables them to be connected together as required.

breakpoint A point in a program where normal operation is suspended automatically, for testing purposes. Breakpoints are specified by the programmer so that interim results of processing can be inspected. The program can then be restarted and will continue running normally.

bridge A means of connecting communications networks (and particularly local area networks) on a large site or across different sites. A bridge operates by 'learning' which of the addresses contained in messages circulating on the networks refer to local addresses and which do not. It then regenerates the latter on the relevant network. It does not switch or interfere with messages in any way, apart from what is necessary to make efficient use of the communications link between the networks and to detect and correct transmission errors that occur on this link. Contrast with *gateway* and *router*, both of which play a more active role, and with *repeater*, which regenerates electrical signals rather than messages.

In terms of the OSI reference model a bridge operates at level 2, the data link layer.

British Approvals Board for Telecommunications (BABT) A UK government agency responsible for testing and approving equipment for connection to public telecommunications services.

British Computer Society (BCS) The UK professional association for people involved with computing.

broadband In general, a high-speed transmission channel. Used particularly to describe local area networks that divide transmission capacity up into a number of channels that can be used concurrently, normally using frequency division multiplexing. Contrast with *baseband* local area networks that encode the signal directly on to the transmission medium.

brouter A piece of network equipment that combines the functions of a *bridge* and a *router*.

browse Search through information in a free-and-easy way, rather than by identifying each item to be viewed explicitly beforehand. By analogy with how people read books, browsing is the equivalent of flicking through the book picking up items that catch the eye, rather than using the index to find items on particular pages. Browsing is a particular feature of hypertext.

BSC Abbreviation of *binary synchronous communications*.

BSP Abbreviation of *business systems planning*.

bubble jet A technology used in high-quality printers, and particularly colour printers. Compare *ink jet printer*.

bubble memory A form of mass storage, manufactured in chip form, but that does not use semiconductor principles. Instead, it relies on the local variations (the 'bubbles') created in thin films of certain magnetic materials, such as garnet. Data in a bubble memory may be thought of as circulating round and round a closed loop within the device. The data passes a single 'window' at regular intervals, at which point it may be read or written. Access is therefore much slower than to semiconductor random access memory (RAM), and closer in speed to that of disk storage. At first, it was believed that bubble memory would prove cheaper than small disk drives, with the added advantage of being smaller and more reliable, but so far it has failed to gain market acceptance.

buffer As a verb, store data temporarily in preparation for transfer or processing. As a noun, an area of memory used for that purpose. With mass storage devices, for example, incoming blocks of records are normally read into buffers, then records are passed across to applications programs one by one for processing. In the outwards direction, records presented by applications programs for output are moved into a buffer, and are only physically written to the storage device when a block is full or no more buffers are available. Buffers are used because it is convenient for an applications program to process records one at a time, but would make very inefficient use of mass storage space if records were also held physically separate on the mass storage device.

See also *double buffering*.

bug A fault in software or hardware, usually minor but often extremely frustrating or damaging, that causes a computer system or program to behave in a way other than how the designer intended. Hence 'debug', meaning to track down and eliminate faults.

bulk storage See *mass storage*.

bulletin board The electronic equivalent of an ordinary bulletin board on which anyone can put up information for others to read. It consists of a computer system shared by a geographically scattered group of users to exchange information electronically. Some general-purpose bulletin boards are operated by commercial service companies, such as Compuserve and Delphi in the US, but many are operated (often on a shoestring budget) by common interest groups, such as people using a particular type of personal computer or interested in a particular hobby or political issue.

Bulletin boards are normally accessed by dialling up from a personal computer over the telephone network.

bundle To sell more than one component of a computer system at a single inclusive price, and not make them available separately. This is a marketing tactic that has been used by manufacturers of complete hardware-plus-software systems to head off competition from specialist manufacturers. For example, by quoting a single price for the central processor and the operating software, they can make it prohibitively expensive for customers to buy the central processor alone from a competitor, because they must then pay the bundled price in addition to obtain the operating software.

burn See *blow*.

burst mode Describes a transfer of information that takes place in a single high-speed burst, rather than a character at a time.

bus In general terms, a channel along which signals travel from one of several sources to one of several destinations. Specifically:

(1) In small computer systems such as minicomputers and personal computers, a set of parallel electronic pathways that carry information between the central components, i. e. those under the direct control of the processor, such as memory and peripheral interfaces, not the peripherals themselves, which are connected via interfaces. The bus for the original IBM PC, for example, consists of 62 lines. It has eight power or ground lines, 20 address lines for memory or port addresses, eight data lines to carry data, and the remaining 26 carry control signals such as interrupt requests and timing pulses.

The key characteristics of a bus are its speed, normally expressed in megaHertz (MHz), and its width, i.e. the number of bits it can carry at once. The original IBM PC bus carries eight, and the bus in a modern personal computer typically carries 32.

(2) In a network, a series of connectors linking devices together – see *bus topology*.

bus topology A topology, used both for local area networks and for the central components of small computer systems, in which all devices are attached directly to a bus – a single continuous channel along which all signals travel to all devices attached to it.

business graphics Graphics used to present relatively simple statistical information, such as is used to support business decisions. Business graphics software can usually present the base data in a variety of formats, including pie, bar and area charts, line graphs and scatter diagrams.

business microcomputer See *personal computer*.

business systems planning (BSP) A method of analyzing the activities of an organisation in order to identify requirements for computer-based systems, originated by IBM. Often abbreviated to BSP. It is a rigorous, formal method which requires managers in those parts of the organisation under study to identify the items of data used by the various business processes. It has been widely used by large organisations to plan their information systems activities, but has recently been criticised as being too oriented towards traditional data processing. It also makes considerable demands on the time and the understanding of the managers involved in the planning exercise.

Compare *critical success factors*; *stages of growth*.

busy hour A continuous one-hour period in which the traffic on a communications network is normally at its highest level during the day.

button (1) A small pad or key on a mouse that can be depressed to send a signal to the computer system to which it is attached.

(2) A small pushbutton-like image on a display screen, labelled with descriptive text such as 'OK' or 'Cancel'. The user activates the button, such as by moving the cursor over it with the mouse and clicking the mouse button (see above), to request, confirm or cancel an action by the computer.

byte A group of eight bits treated as a unit. Normally used either to represent a single character or two binary coded decimal digits. The memory of many computer systems, and particularly of personal computers, is organised and addressed in terms of bytes.

byte ordering The position that bytes occupy in a word of memory. Some computer processors place the first byte (of the two that occupy a word) in the upper part of each word and others place it in the lower part. This is significant if programs or data are to be moved between computers of different types, since operations that rely on finding bytes ordered in a particular way will no longer work correctly.

C

C A high-level programming language widely used to develop systems software and applications packages for small computers, including personal computers. C is a block-structured language, and produces programs which can easily be moved from one type of computer system to another. Although a powerful high-level language, it also allows programmers to access low-level functions. These two features account for its popularity for developing personal computer software. An enhanced version of C has been produced, called C++, which supports object-oriented programming.

cable network A network linking homes, usually by coaxial cable, to provide TV programming of superior quality or variety to that available by normal broadcast channels. Cable networks can also be used to carry data, such as via videotex technology.

cable TV A network linking television sets to a 'head end' broadcasting station, usually by coaxial cable. The 'head end' station generates the broadcasting signals and sends them out over the cable.

cache memory An area of memory where a computer system stores copies of the sections of program and data recently retrieved from direct access storage such as disk. If a program attempts to reference these, then the stored copies are used to save the time involved in accessing mass storage. An algorithm based on frequency and last time of use is used to decide when copies in the cache memory should be replaced. This means that sections of program that are accessed frequently, such as operating system overlays, remain permanently in memory.

CAD Abbreviation of *computer aided design*.

CADCAM Abbreviation of *Computer Aided Design Computer Aided Manufacturing*, meaning systems integrating those two concepts.

CAE Abbreviation of *common applications environment.*

CAE Abbreviation of *computer aided engineering.*

CAI Abbreviation of *computer assisted instruction.*

CAL Abbreviation of *computer assisted learning.*

call (1) A term used to describe the process of communication between non-voice devices such as computers and terminals, particularly over public telephone or data networks, by analogy with telephone calls. A non-voice call consists of three phases: (i) call establishment, in which connection is established between the devices; (ii) the data transfer or exchange of messages; (iii) call clearing, in which the devices disengage in an orderly manner.

(2) A program instruction or statement that is used to enter a subroutine. The address of the instruction or statement following the call (known as the return address) is stored automatically when control is transferred to the subroutine. When the subroutine has completed its task, program execution continues from the instruction or statement following the call.

call accepted packet A special packet sent to the network by a terminal on a packet-switching network (the called terminal), indicating that it accepts the call request received from another terminal (the calling terminal). As a result, a virtual circuit is set up by the network between the two terminals, and a call connected packet is sent to the calling terminal.

call blocking A condition which arises in a fully loaded telephone exchange or a similar switching system, when no further new calls can be accepted. Such systems are normally configured so that the probability of call blocking occurring in the busy hour is around 0.01.

call clearing The phase that concludes a call between devices using a communications network, in which the devices disengage in an orderly manner.

call connected packet A special packet sent to the calling terminal by a packet-switching network, indicating that the call has been accepted by the called terminal and that a virtual circuit has been set up by the network between the two terminals.

call establishment The phase that initiates a call between devices using a communications network, in which the devices establish a connection between one another.
See also *connection-oriented network service.*

call request packet A special packet sent to another terminal (the called terminal) by a terminal on a packet-switching network (the calling terminal), indicating that it wishes to establish a call between the two terminals.

call setup See *call establishment*.

called terminal The device that responds to a request from another device (the calling terminal) to establish a call, for an exchange of messages or for data transfer.

calling rate A measure of the use a device makes of a communications network, normally expressed as a calling rate of so many calls per hour of so many minutes duration.

calling sequence The sequence of program instructions or statements necessary to call a subroutine or a procedure.

calling terminal The device that initiates a call to another device (the called terminal) by sending a call request.

CAM Abbreviation of *computer assisted manufacturing*.

Cambridge Ring A local area network with a ring topology, developed by computer scientists at Cambridge University. It uses twisted pair, coaxial or fibre optic cables to carry data at 10 megabits per second, and a token passing protocol to control traffic.

CAN Abbreviation of *cancel character*.

cancel character (CAN) A control character included in a message that is transmitted, indicating to the receiving device that it should ignore the data preceding the character in the message.

CAR Abbreviation of *computer assisted retrieval*.

card See *expansion card*.

card punch An output device, now obsolete, that encodes data into cards by punching holes in them.

card reader An input device that reads cards and transfers the data held on them into computer memory for processing. Card readers were widely used in the 1960s and 1970s to read data and source programs punched into 80-column cards. They are now used on pocket-sized computers to read and write integrated circuit memory cards. These can be used to transfer data between the portable computer and larger ones used at home or in the office.

carriage return character (CR) A control character that tells the receiving device to move the print head (in the case of a printer) or the cursor (in the case of a visual display terminal) to the beginning of the next line. (Some older devices such as teletypewriters, move the head to the beginning of the current line – these devices only advance to the next line when they receive a line feed character.) It is generated by pressing the RETURN key on most computer and terminal keyboards. It is also used as a record delimiter in ASCII files – fields within each record are separated by comma or TAB characters, and each record within the file ends with a carriage return character.

carrier See *carrier signal*.

carrier See *common carrier*.

carrier sense multiple access (CSMA) A method of controlling the traffic sharing a single channel of a local area network. Often abbreviated to CSMA. The most widely used version is known as carrier sense multiple access with collision detection (CSMA-CD). CSMA-CD is used on local area networks such as Ethernet. It requires each device on the network to 'listen' constantly to traffic using it and only attempt to send a message when there is no traffic. If two devices send at exactly the same time – a collision – each of them waits a random period of time before trying again. This method works very well when traffic on the network is within its carrying capacity, but performance declines rapidly as the load reaches a critical point.
    See also *contention*.
    Compare *token passing*.

carrier signal A signal of a chosen frequency, generated to carry a message signal. The message signal is imprinted on the carrier signal by a process known as modulation.
    Compare *baseband*.

carry register A special register used by the arithmetic/logic unit of a processor to hold the arithmetic carry for calculations involving more than one word.

cartridge tape A magnetic tape housed in a cartridge. Normally used to refer to computer industry-standard 1/2 inch tapes. The advantage of the cartridge is that the operator can load the tape without needing to touch the storage medium itself.

CASE Abbreviation of *computer-aided software engineering*.

case-sensitive Of operations on text information, such as keyword searches or sorts, that treat upper case letters as different from lower case.

CASE statement A statement used in some high-level languages to branch to a series of options based on the value of a variable.

catalogue See *directory.*

cathode ray tube (CRT) The component of a television set and of most visual display devices that creates the image seen on the screen. Normally abbreviated to CRT and sometimes used as synonymous with visual display, as in CRT terminal. It works by directing a beam of electrons at the surface of the screen, controlled by an electronic lens.
   Compare *liquid crystal display.*

CATV Abbreviation of *community antenna television.*

CAV Abbreviation of *constant angular velocity.*

CBMS Abbreviation of *computer-based message system.*

CBT Abbreviation of *computer-based training.*

CCD Abbreviation of *charge-coupled device.*

CCITT Abbreviation of Comité Consultatif International de Téléphonie et de Télégraphie, a committee set up by the world's telecommunications authorities to establish standards for telegraphy, telephony and, more recently, data communication. It is part of the International Telecommunications Union (ITU), a treaty organisation of the United Nations. For data communication, CCITT has established two main series of standards recommendations, the V series dealing with communication over public telephone networks and the X series dealing with communication over special-purpose data networks.

CCTA Abbreviation of *Central Computing and Telecommunications Agency.*

CD-A The standard used for digital recording of information on ordinary music compact disks.
   Compare *CD-ROM.*

CD-I A standard for storing digital information on compact disk originated by Philips, abbreviated from compact disk interactive. The standard describes both the storage media and the hardware that retrieves information from it. CD-I devices are consumer products sold as 'players' that can be attached to a television set. The CD-I player presents voice, sound effects, and still and moving video pictures, as well as massive data files, all under interactive control of the user. A competing standard known as *digital video interactive* (DVI) has been developed by RCA and is owned by the chip manufacturer, Intel. Contrast also with *CD-ROM* disks, which can only be read by a computer.

CD-ROM Abbreviation of 'compact disk read-only memory', a format for storing computer-readable data on a compact disk (defined by ISO standard 9660). The same disk that is used for audio can hold between 550 and 630 million characters of information, depending on the format used, each encoded with 8 bits. These bits are permanently recorded on the disk as tiny bubbles burned in with laser optics, and it is this which makes it a read-only memory. The storage capacity represented by 550 million characters is the equivalent of about 250,000 printed pages or 300 substantial textbooks. Alternatively, a disk could hold, digitally encoded, 2000 high-resolution colour images or 20 hours of recorded speech.

Like vinyl long-playing records, CD-ROM disks can be 'mastered' and pressed in quantity at between $5 and $10 per copy. Unlike vinyl LPs, though, CDs are read by scanning, so they do not wear out through physical contact.

The cost of preparing a CD-ROM master is in the $1,000s – much less than the cost of printing an equivalent volume of material.

There is an enhanced version of the CD-ROM format, known as CD-ROM XA (extended architecture) and only playable on what is known as a mode 2 drive. This adds some of the features of CD-I, with the idea that a single publication designed for CD will play on both a CD-ROM and CD-I player.

See also *CD-WO*; *photoCD*.

Compare *WORM*.

CD-WO Abbreviation of compact disk write once, a special kind of CD-ROM drive that can be used to originate a disk, using a special 'gold disk'.

CDOS Abbreviation of *concurrent DOS*.

Ceefax The trade name for the BBC's teletext service.

cell relay An architecture for high-speed digital communications networks that enables voice, data and video to be sent in parallel. It is the key technology underlying the ISDN (integrated services digital network) services to be offered by many telecommunications carriers.

Compare *frame relay*.

More detail *asynchronous transfer mode*.

cellular radio A technology that supports mobile telephone services. The key to the technology is a central computerised automatic switching system within a metropolitan or rural area in which the service operates. The total area is divided into cells, each of which contains a base station transmitting only to radio telephones within its boundaries. As the user moves out of one cell and into the next, the conversation is switched to the next base station automatically. This arrangement allows frequencies to be reused, which means that more simultaneous calls can be handled within a given area.

Central Computing and Telecommunications Agency (CCTA) The unit responsible for overseeing UK Government computing. Its role is to advise Government departments on methods and to set standards for such things as procurement. At present it is part of the Treasury.

central processing unit (CPU) A term formerly used to describe the processor of a large computer system, in the days when the processor was a substantial unit, usually occupying its own separate cabinet.

central processor See *processor*.

CGA Abbreviation of *colour graphics adaptor*.

CGM Abbreviation of *computer graphics metafile*.

chain A method of organising a sequence of related data items, such as the records within a data file, by pointers in each record indicating the next (and sometimes also the previous) item in sequence.

champion See *project champion*.

channel In general terms, a path for electrical transmission between two or more points. Used particularly to refer to the logically separate paths into which transmission lines and communications links can be divided so that a number of devices, terminals for example, can use them concurrently.

Also used to refer to the components of large computer systems that carry data between peripherals and memory (small systems usually have individual interfaces connected to the bus, rather than channels). These channels come in two main types – selector channels that carry bursts of data from high-speed peripherals such as disk drives; and multiplexor channels that carry the traffic from a number of slower peripherals such as printers or terminals.

character A letter, numeral or special symbol, forming part of a character set.

character code A bit pattern that is used to represent a character in the memory or on the mass storage devices of a computer system, or in a message sent to a visual display terminal, a printer or some other device. Also used to mean a complete set of bit patterns representing the full range of characters available – see, for example, ASCII or EDCDIC.

character generator The component of a visual display device that converts the codes used to represent characters into the shapes of letters and numbers displayed on the screen.

character keys Any of the keys on a keyboard – letters, numbers, punctuation marks, symbols – that produce output on screen or printer when pressed. This is in distinction to keys such as Shift, Control, Caps Lock, etc. that modify the effect of character keys but do not of themselves produce output.

character mode terminal Used of terminals that send messages a character at a time, rather than assembling complete messages for transmission.
   Compare *packet mode terminal*.

character printer A printer that prints a character at a time, rather than a line or a page. There are two main types of character printer, both of which form characters by striking an inked ribbon against the paper: (1) matrix printers form each character out of a grid of dots, typically 9 dots by 9; (2) daisywheel printers have a rotating metal or plastic wheel, each spoke of which has the raised shape of a character – one of these is brought under the print hammer and forced against the ribbon.
   Compare *line printer*; *page printer*.

character set The set of characters, each represented by a different bit pattern or code, that a particular computer system or program accepts as valid and/or that a standards-making body defines as valid. For example, the ASCII character set ratified by ISO and used by most personal computers, or the EBCDIC character set defined by IBM for its Series 360 and subsequent ranges of general-purpose computers.

character string See *string*.

characters per second (cps) A measure of the speed of character printers or other character-oriented devices. Sometimes used also as a measure of the speed of transmission of data.

charge-coupled device (CCD) A technology consisting of arrays of capacitors that can be charged up, such as by exposing them to light. The pattern of charging can then be read out. Often abbreviated to CCD. The technology is used for document scanners, in which CCDs are used to measure the lightness or darkness of each pixel of the document and to assign a numerical value to them.
   CCD technology was also seen as suitable for mass storage systems, but has yet to make a commercial breakthrough. It has cost and access time characteristics lying between those of disk and semiconductor memories (RAM). Unlike RAM devices, data held in a given CCD component can only be read out serially, but a storage unit can be subdivided into a number of such components, each of which can be addressed individually. This gives the effect of direct access such as on disk drives.

cheapernet See *thin Ethernet*.

check digit An extra digit added to a numeric field, such as a credit card number, to minimise the possibility of incorrect codes being used by mistake. A check digit is normally calculated by finding the remainder when the original number is divided by a fixed number. The same calculation is carried out by the computer whenever the code is entered, and the code is rejected if the check digit is found to be incorrect.

checkpoint Details of the status of a program and of its peripherals and terminals, recorded on mass storage at a chosen point in its execution. Should the program subsequently fail, the checkpoint is used to reconstruct its status at the time so that it can be restarted from a known point.
See also *cold restart*.

checksum A simple error detecting mechanism. It consists in generating a number by adding together all the individual digits of a field or all the individual characters in a message and appending the result, known as the checksum, to them. Any system receiving this information does the same and, on finding a mismatch, knows that the message has been changed or corrupted.

chicken-and-egg loop A loop in an expert system or in logic programming, caused by conditions in two different rules depending on one another.

chief information officer (CIO) A senior manager with responsibility for information and information systems within an organisation. The idea of the chief information officer or CIO arose from the concept of information resource management or 'information as a corporate resource' – the CIO's job would be to put the concept into practice. 'Information as a resource' was a broadening of the database concept, which saw data files as a model of the data an organisation required in order to operate.

chip See *integrated circuit*.

chip card See *smart card*.

CIM Abbreviation of *computer-integrated manufacturing*.

CIO Abbreviation of *chief information officer*.

circuit The physical connection of equipment between two given points, so that information can be exchanged between them.
See also *virtual circuit*.

circuit-switched Of a network that uses circuit-switching techniques.

circuit switching A form of switching in which a physical transmission path is established between two devices exchanging information. The two devices have full and exclusive use of that transmission path until their exchange terminates. Telephone networks normally use circuit switching, and some types of data network do also.

See also *fast circuit switching*.

Compare *message switching*; *packet switching*.

circular shift See *logical shift*.

CISC Abbreviation of *complex instruction set computer*.

CIT Abbreviation of *computer-integrated telephony*.

class hierarchy A mechanism which allows universal truths – generic attributes of a particular class of objects – to be defined only once, and to be inherited implicitly by members of the class. Class hierarchies are used within databases, where an entity in the database belonging to a particular class automatically inherits the attributes of that class (see *attribute inheritance*). They are also used in object-oriented programming, where program 'objects' in a particular class inherit the behaviour of that class. The class, then, is the template from which an object is derived, defining what data and functions are associated with it.

See also *generalisation hierarchy*.

clear down Terminate a call, in the same way that telephone calls are terminated by replacing the handset.

clear request packet A special packet sent by either of the terminals engaged in a call on a packet-switching network, indicating that it wishes to clear down the call between the two terminals.

clear to send (CTS) A signal used to control data flow between two directly connected devices, such as a computer and a modem. The sending device raises a ready to send (RTS) signal when it has data to send. The receiving device raises a clear to send signal when it is ready to receive a transfer of data, such as when a memory buffer is available.

client A device or an applications program that uses the services offered by a server.

client-server An arrangement for sharing work between equipment cooperating on a processing task. The server provides a given range of services to its clients, which might be personal computers connected to it via a local area network. Whenever they require a service, such as retrieval of specified data from a database, they send in a request to the server which acts on it and returns the results. The key aspect of this arrangement is that server and client carry out their respective tasks independently of one another, whereas in a conventional arrangement such as a teleprocessing system the terminals operate as an extension of the main computer and are controlled by it.

See also *X Windows*.
Compare *master-slave*.

CLNS Abbreviation of *connectionless network service*.

clock circuit A circuit used to synchronise the operation of a computer processor. It emits electronic pulses at fixed intervals. These are used to control the timing of all circuits within the processor, and thus determine the speed of processing (also known as the *clock rate*).

clock rate See *clock circuit*.

clone A computer system that operates in the same manner as another, so that it can run the same software without alteration. The term is most widely used to describe personal computers that are designed to be compatible with the IBM PC.

CLOSE file A command issued by a program to indicate that processing of a file is complete. On receiving the command, the operating system makes sure that all operations affecting the file have been completed, such as for example writing any updated records buffered in memory to the file; releases any system resources allocated to the file, such as memory buffers; then makes the file available for processing by other programs.

closed user group (CUG) An arrangement which permits terminals using a shared service, such as a public packet switching or videotex service, only to make calls to or receive calls from other terminals belonging to the group. The membership of the group is specified to the service operator in advance.

cluster A group of devices, such as terminals or processors, that are located close together and can be addressed both as a group and individually. Clusters of terminals are linked to the processor via a single cluster controller.

cluster controller A device that controls a cluster of terminals. A single communications link runs to the cluster controller, and this is connected in turn to the terminals in the cluster, typically visual display terminals located within a limited distance on the same site. Cluster controllers include logic that would otherwise have to be built into every terminal, thus reducing overall cost, and they also enable a number of terminals to share a single communications link.

CLUT Abbreviation of *colour look up table*.

CLV Abbreviation of *constant linear velocity*.

CMIP Abbreviation of *common management information protocol*.

CMOS Abbreviation of *complementary metal oxide semiconductor.*

CMYK Abbreviation used to refer to the technique used to print high-quality colour material such as magazines. It refers to the four coloured inks that are combined to produce a complete range of colours. They are cyan (a kind of turquoise), magenta (a cerise pink), yellow and key (black). Some desktop publishing software packages can produce CMYK colour separations to be used in the colour printing process. Contrast with *RGB* (red-green-blue) – the three colours used to produce colour images on computer displays and TV screens.

CNC Abbreviation of *computer numerical control.*

coaxial cable A transmission medium widely used for connecting terminals and small computers on-site, including in local area networks. Also used domestically to connect television aerials. It consists of a pair of conductors, a core within an outer sheath, separated by insulating material. The outer conductor serves to reduce electrical interference. Coaxial cables are capable of bit rates up to 60 megabits per second.

COBOL Contraction of COmmon Business-Oriented Language, a high-level language widely used in business for commercial applications. It was designed in 1959 to meet the needs of the US Department of Defense by Dr Grace Hopper. It is particularly oriented towards processing of data files. In theory, COBOL is a standard language, available on a

---

Structure of a COBOL program

COBOL programs are divided into four 'Divisions'.
(1) The Identification Division contains the name and other identifying details of the program.
(2) The Environment Division specifies the characteristics of the computer system required to compile and run the program, such as size of memory, number and type of peripherals needed.
(3) The Data Division associates each file that is to be processed with a peripheral, and shows the layout of the records held in the files and of any work areas and constants that the program will use. Each field is given an identifying name or label. These labels are used in the Procedure Division to reference the field concerned.
(4) The Procedure Division contains the language statements that solve the processing problem. Each statement consists of reserved words or phrases standing for particular operations, such as READ, ADD ... TO, AT END GO TO; conjunctions and logical operators such as IF, AND, LESS THAN; and labels standing for fields already defined in the Data Division.

---

wide range of different computer systems, but in practice source programs cannot always be moved from one type of computer to another without modification.

The main advantage of COBOL is that it is relatively easy to learn and to understand, since it uses English-like constructions and keywords. This means that programs can be written, tested and maintained easily. Its drawback is that it is long-winded – the Data Division in particular is verbose and repetitive – and lacks flexibility. It is in its element dealing with the type of application it was designed for – commercial batch processing applications handling multiple files – but unwieldy when called on to handle applications that stray far outside those limits.

COCOM Abbreviation of *Committee for Co-ordinating Multilateral Trade*, a committee set up by the US Government on which the world's leading industrial nations are represented. It is intended to ensure that advanced technology likely to be of military use, including many information technology products, is not traded carelessly.

CODASYL Abbreviation of COnference on Data SYstems Languages, a body that coordinates the development of the COBOL programming language and related languages, and particularly those used to build and use databases based on the network model.

code (1) As a verb, to write a program in a computer programming language. As a noun, the result of doing so.

(2) A set of rules for interpreting a series of bit patterns. Thus machine code is a set of bit patterns which represent different instructions for a particular computer, and ASCII code is a set of bit patterns representing alphabetic and other characters.

code conversion Conversion of a series of characters by substituting the codes from another character set for the codes in the character set that was used originally.

codec See *network interface unit*.

coding The process of working out and writing the language statements that make up a source program.

cold restart A restart carried out by returning to a previous checkpoint or to the beginning of the run that was in process when the program halted, and repeating the processing performed since then. Contrast with *warm restart*, which is where the program can be restarted without going back to an earlier point in this way.

collate See *sort*.

collating sequence The sequence into which keys are placed when they are sorted. This sequence is affected primarily by the binary values assigned to different characters in the character set used for the key, although the program carrying out the sort may override this. Some sort routines, for example, treat upper and lower case alphabetic characters as having the same value.

collision Where more than one device sharing a transmission channel attempts to send at the same time. Used particularly of local area networks.
  See also *carrier sense multiple access*.

colour gamut The number of colours that a display screen can reproduce. This may run into millions, although the screen will normally only be able to display a limited number of these – say 256 – at any one time, which the user selects via the operating software – see *colour look up table*.

colour graphics adaptor (CGA) A basic standard for the display of information on the screen of a personal computer, and also a plug-in card that implements the standard. The standard was introduced along with IBM's original PC in 1981. It provides for both text and graphics, implemented in two separate modes. In text mode, characters are constructed from a 8 x 8 matrix of dots. In graphics mode, information can be displayed in up to four colours in a screen resolution of 320 x 200 pixels, or in monochrome at a resolution of 640 x 200 pixels.

colour look up table (CLUT) A method of encoding the information necessary to generate colour pictures on a computer display screen. Definitions of a number of colour shades are held in a table, and values representing positions in that table are held for each pixel on the display screen. Typically 4 or 8 bits are used, permitting 16 or 256 different colours respectively.

colour management system Software that helps to match the colours produced on different devices, such as those shown on a computer display screen with those produced by a colour printer or a slide recorder.

colour purity A measure of the quality of a colour display, reflecting its ability to display pure white.
  See also *colour uniformity; convergence; grey linearity*.

colour separation See *separation*.

colour uniformity A measure of the quality of a colour display, reflecting how uniformly it displays white over the entire screen. Many displays have *warm spot*s where colour or brightness levels vary.
  See also *colour purity; convergence; grey linearity*.

COM Abbreviation of *computer output microfilm*.

combined fact-rule uncertainty The degree of uncertainty of a conclusion drawn by an expert system, achieved by combining the confidence factors for the facts and the rules included in the knowledge base.

command and control system A computer-based system used to coordinate military or paramilitary operations, such as a police force.

command-driven Of a program that is controlled by keying in a series of commands rather than, for example, by selecting options from menus.
  Compare *menu-driven*.

command line interface A general term for a user interface (the mechanisms that enable people to converse with computers) that is driven by commands that the user types in at the keyboard. Most of today's personal computers – those running under the MS-DOS and PC-DOS operating systems – use such an interface.
  Compare *graphical user interface*.

comment A program statement ignored by the compiler other than to print it on the program listing. It is used by the programmer to give information about the operation of a program to people reading it.

commercial data processing Data processing to support administrative work and management, rather than scientific work or manufacturing processes. Systems supporting commercial data processing need to be able to handle large data files held on mass storage, and networks of terminals used for data entry and enquiry.

common applications environment (CAE) A set of standards for the software components used to develop and interconnect computer applications defined by the X/Open group of computer suppliers.

common carrier An organisation that supplies telecommunications services to the public in general or sometimes to specific classes of the public. In most countries, common carriers are subject to state regulation and operate as a monopoly or near-monopoly, although there is a widespread move toward deregulation of telecommunications services.

common control Used to describe telephone exchanges in which switching is controlled by a digital computer, rather than by signalling units associated with each line entering the exchange.

common management information protocol (CMIP) A standard protocol for the interchange of network management information within computer networks. It forms part of the open systems interconnection (OSI) standards being formulated by ISO.

46

Compare *simple network management protocol*.

communicating word processors See *electronic mail*.

communications interface A hardware component of a computer system or terminal which enables it to be connected to a transmission line (normally via a modem or network interface unit). Sometimes used also to mean a definition of how the interface should work.

communications link A means of connecting terminals and computer systems so that messages or other forms of information can be transferred between them. A communications link may comprise a number of transmission lines and/or communications control devices and/or public telecommunications networks, but the devices at each end of the link need not be aware of this. They expect to post information at one end of the link and have it received at the other.

communications network A collection of telecommunications equipment and transmission lines, used to interconnect devices (computers, terminals, telephones, fax machines, etc.) at different locations so that they can exchange information.

communications processor See *front-end processor*.

communications server See *server*.

communications standard See *standard*.

community antenna television (CATV) Television signals distributed to sets within a given area, usually via coaxial cable, from a shared aerial.

compact disk Compact disk technology, now accepted as the standard for digitally encoded audio (known as *CD-A*), was developed by Philips and Sony, using a 12 cm. format for the disk. The same disk that is used for audio can also be used to hold digital information in a number of other forms – see in particular *CD-I* and *CD-ROM*.

compact disk interactive See *CD-I*.

compaction See *memory compaction*.

compatibility The ability of one component of a computer-based system to work with another. This can apply at a number of different levels:
  (1) *Program* compatibility means that a program that runs on one type of machine will also run on another.
  (2) *Language* compatibility means that a program written in that language can be

compiled and run (also known as *portable*).

(3) *Plug* compatibility means that a hardware component can replace another and plug into the same interface socket.

It is important to recognise that there are degrees of compatibility. Thus a program may run on another computer that is nominally compatible, but may perform so badly that it is unusable. By analogy, two people may be compatible when it comes to casual social contact, but quite incompatible when it comes to marriage.

Compatibility is an issue not only with products from different suppliers, but also with upgraded versions of the same product – see *downwards compatible* and *upwards compatible*.

Compare *portable*.

compile time The time when a program is compiled. The term is particularly used to describe actions or decisions taken at this stage rather than, for example, when the program is being run.

Compare *run time*.

compiler A program that translates a source program into an equivalent program in machine language (an object program), so that it can be run by a computer. To produce the object program, compilers have to: (1) translate each statement into machine language equivalents; (2) incorporate into the object program any library subroutines required (explicitly or implicitly) to run the program; (3) supply the interconnecting links between components of the program.

A compiler for a powerful language can be a very complex program and may complete its task in more than one stage.

Compare *interpreter*.

complement The normal method of representing negative numbers within computer systems. The complement of a number is obtained by inverting each bit. In noughts complement (normally used by digital computers), one is then added to the result. The lefthand bit of a word is usually reserved to represent the sign, zero meaning positive. Thus an 8-bit data field can hold values between 127 (01111111) and -128 (10000000).

complementary metal oxide semiconductor (CMOS) An integrated circuit technology consisting of pairs of transistors of opposite type (i.e. n-p-n and p-n-p). It is used for applications where low power consumption and heat dissipation is required, such as in portable computers.

complex instruction set computer (CISC) A computer with a conventional instruction set, corresponding directly with the instructions in an assembly language.

Compare *reduced instruction set computer*.

component video A video signal used by professional videotape systems, allowing greater quality than the *composite video* signal produced by many domestic VCRs.

composite video A video signal combining colour picture (RGB) and timing information, as produced by many domestic VCRs.

Compare *component video*.

compound document An electronic document composed of content of different types, such as text in character form as well as graphics.

compound statement A group of program statements that are treated as a unit by the compiler. For example, a compound statement may be used for the THEN or ELSE clauses of an IF statement.

computer A machine that can accept a series of instructions (a program), and can then interpret these instructions while accepting and processing data, in order to generate new information or to control another machine or process. Computers fall into two main categories – analog computers, used mainly in scientific and industrial applications; and

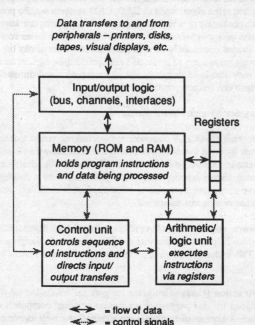

Data transfers to and from peripherals – printers, disks, tapes, visual displays, etc.

Input/output logic (bus, channels, interfaces)

Registers

Memory (ROM and RAM) *holds program instructions and data being processed*

Control unit *controls sequence of instructions and directs input/output transfers*

Arithmetic/logic unit *executes instructions via registers*

⟷ = flow of data
⟵⋯⟶ = control signals

digital computers, which are used in similar applications and a wide range of other commercial, industrial and military applications. Digital computers are the driving force behind the advance of information technology. They have developed through a series of 'generations', each using a more powerful base technology than the previous. The first generation of computers used thermionic valves, and the second used transistors. The digital computers being delivered today belong to the fourth generation, using large (LSI) or very large scale (VLSI) integrated circuits.

The term 'fifth generation computer' is used in a special sense, however. It is used to describe machines, still under development, that will be controlled and operated by means of artificial intelligence techniques rather than by conventional programs.

Strictly speaking, the computer is only the machine that does the work of processing information, and its main components are shown in the figure. For most practical purposes it must be complemented with peripherals and terminals that people can use to get the information into and out of it. Together with these devices, it forms a computer system, although the term 'computer' is sometimes used in this sense also.

computer aided design (CAD) The use of computers to create the geometric detail and labelling involved in product and architectural design and drafting. Also known as *computer aided drafting* and often abbreviated to CAD. CAD systems usually consist of one or more workstations, consisting of a keyboard, a high-resolution display screen, and either a light pen, a mouse or a graphics tablet. The workstation may be standalone or may be connected to a shared computer with disk storage. The software in the workstation and shared computer helps users to create and manipulate drawings, in either two or three dimensions. It may also hold details of the characteristics and dimensions of parts and components which can be incorporated into drawings.

computer aided drafting See *computer aided design*.

computer aided engineering (CAE) A computer system linking product design and manufacturing, most often found in the electronics industry. Also known as CADCAM. It might consist, for example, of CAD workstations that supported the design of integrated circuits and printed circuit boards, linked to a processor with a database that provides output to drive computer numerically-controlled (CNC) drills, automatic component insertion, and automatic test equipment.

computer-aided software engineering (CASE) A general term for sophisticated computer-based tools which help systems analysts and programmers to design and build applications programs. This includes analyst/designer workbenches and system building tools.

computer assisted instruction (CAI) Computer support for education or training. Software packages designed for this purpose either run on personal computers or on a shared processor to which a number of terminals are attached. As well as presenting a body of knowledge, displaying text or animated graphics on the screen, perhaps using synthetic speech as well, they can set simple exercises to verify that the student has understood the material.

computer assisted learning (CAL) A general term for the use of computers to help individuals or groups in learning activities such as education or training. It includes well-established technologies such as computer-based training (CBT) and the more recent interactive learning systems (ILS).

computer assisted manufacturing (CAM) A general term for computer support for the manufacturing process. It is often bracketed with computer aided design, in the abbreviation CADCAM.

computer assisted retrieval (CAR) Computer support for retrieval of information stored on microfilm. It usually consists of a computer system that maintains an index on disk of documents stored on microfiche or in a similar form. The index can be interrogated from data terminals and the system either identifies the fiche required for manual retrieval, or may be linked to units that retrieve the fiche automatically.

computer audit An auditing process applied to computer systems. A computer audit may cover any or all of the following: establishing that internal and external auditors are satisfied with the audit trail that systems provide; verifying that systems are operating according to specification and are giving the results they are supposed to give; checking that systems have adequate controls and that security procedures are working properly; checking that adequate security disciplines are being maintained within the data processing department.

computer-based education See *computer assisted instruction*.

computer-based message system (CBMS) Synonymous with electronic mail.

computer-based system A computer system, plus the people and procedures within an organisation that make use of it, interacting to meet some defined goal. In the field of information technology, the computer will be a digital computer and the goal will be to generate and/or to handle information.

computer-based training (CBT) Training using computer-based software tools. Computer-based training is widely used to train end-users and students in the use of personal computer packages. It has the advantage that trainees can practice what they learn as they go along, and can learn at their own pace and at whatever time suits them.

computer centre See *data centre*.

computer conferencing Electronic conferencing by people using personal computers or display terminals, mediated by a shared computer system accessed via a network. Conferences on particular topics are advertised on the shared system, together with an identifying name or code. Participants 'sign on' to conferences they are interested in, and may then display the previous contributions of other participants on their screen and enter their own contributions via the keyboard. These may either be distributed automatically to all conference participants when they next sign on, or be directed just to particular participants. Computer conferences may continue over long periods of time and develop lengthy electronic transcripts.

Compare *video conferencing*.

computer console See *operator console*.

computer department See *information systems department*.

computer fraud A fraud committed by exploiting loopholes in a computer system, or by employing inside knowledge about how a computer system works. Most computer fraud is committed by insiders such as programmers who can attempt to escape detection by covering up their tracks before the fraud is noticed.

computer games Games that run on a (usually personal) computer. Computer games fall into two main categories: arcade games such as Space Invaders and the like, and adventure games where the user has to find a way past various hazards and puzzles.

computer graphics A general term meaning information drawn on a computer screen, rather than displayed as lines of text. It may consist of lines, curves and shapes, entered as freehand drawings (with a mouse or graphics tablet); copied by means of a scanner; or generated by software (for example, bar charts). The reason for making the distinction with lines of text is that terminals designed for data processing can only display text (and a limited range of special characters) in preset positions on the screen – normally 24 lines of 80 characters each – see *80-column display*. Many personal computers operate in this manner, but can also be switched into a graphics mode in which the entire screen is treated as a grid of dots, so that drawings, shapes, etc. can be displayed. Often a plug-in adaptor card is needed to generate these graphics – see, for example, *enhanced graphics adaptor* (EGA) and *video graphics array* (VGA). More advanced machines, such as expensive engineering workstations and also the Commodore Amiga and the Apple Macintosh, have *bit-mapped displays* which handle text and graphics together and with equal ease.

computer graphics metafile (CGM) A standard for the formats used to send two-dimensional computer graphics between computer systems or from a system to a peripheral. It has been approved by ISO as part of open systems interconnection standards (coded ISO 8632) and is supported by most minicomputers.

Compare *encapsulated PostScript format; tagged image file format*.

computer-integrated manufacturing (CIM) Computer-integrated manufacturing (CIM) has many definitions, most of which incorporate the idea of an integrated database of information related to design and manufacturing, and perhaps also administration, so that an order or new design need be input only once for use by every business function. It is associated with an overall business philosophy aimed at closer integration of the various functions that contribute to the manufacturing process.

computer-integrated telephony (CIT) A protocol for the interconnection of computers and private telephone exchanges (PABXs), for applications such as voice messaging. The protocol was originated by Digital Equipment Company (DEC) which is trying to establish it as a standard. Audio media integration standard is a competing standard backed by the US Information Industry Association.

computer literate Applied to people who have enough familiarity with computers to make sensible decisions about them. Sometimes a further distinction is made between computer literate, meaning knowing how to use a computer, and application literate, meaning knowing what can usefully be done with them.

computer network Often used as synonymous with communications network, but sometimes in a narrower sense to mean a communications network used by computer systems, rather than by data terminals and similar devices.

computer numerical control (CNC) Control of machine tools by computers. Since the 1940s machine tools have operated under the control of punched tapes which specify numeric parameters such as position. These are known as numerically controlled (NC) tools. Originally the parameters were calculated by engineers but, with the arrival of small low-cost computers, that task could be taken over by the computer.

computer operator A person whose job is to operate computer systems, acting on instructions supplied by the systems analysts and programmers who produced the programs that are to be run. Operators in large computer installations are expert in getting the best out of the equipment, using the features built into the operating system to do so. They also load storage media such as magnetic tapes and disk packs on to mass storage devices when they are needed; look after peripherals such as printers; and distribute computer output to wherever it is required.

computer output microfilm (COM) Computer 'printout' generated on microfilm rather than on paper.

computer policy See *information policy*.

computer-supported telephony (CST) Applications involving a computer linked to a private telephone exchange (PABX) or an automatic call distribution (ACD) system. The computer supports staff receiving telephone calls by, for example, presenting scripts for them to work through or by displaying automatically information identifying the caller, based on the number they called from.

See also *audiotex*; *interactive voice response*.

computer system A combination of devices, centred round a computer, that are inter-connected and operate in cooperation with one another. A computer system is made up of three resources: (1) hardware, the physical devices that capture, process, store, transmit and display information; (2) software that controls and manages the hardware; (3) data, held on disk or other storage media.

Its main component parts (see figure) are:

- a processor, which controls the system as a whole and processes and transforms information;
- memory, used to hold programs and for short-term storage of information;
- peripherals, such as disk drives and printers, which serve to carry information between the system and its human users, and to store it for processing later;
- input/output logic, controlled by the processor, which carries information between memory and the peripherals.

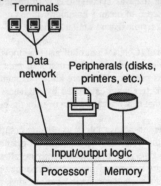

The computer itself is normally regarded as comprising the processor, memory and input/output logic, plus a device used to control it, such as a visual display screen and keyboard on a small computer or a console typewriter on a larger one.

The term is sometimes used to include the people and procedures that use a computer system for a particular purpose, such as to handle a payroll or to control stock levels, but this is more accurately described as a computer-based system. Computer-based systems in turn are one of the technologies, and a very important one, used within information systems. Information systems also use long-established technologies such as

54

paper filing systems, and serve to process and distribute the information that organisations use to plan, monitor and control their activites.

The figure below illustrates this hierarchy – the items listed in the lower part of the boxes are other resources included at this level as well as the information technology resource listed in the upper part.

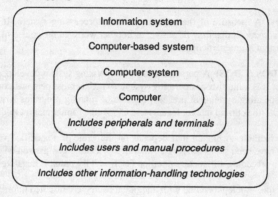

computer vision See *machine vision*.

computerised tomography (CT) The technique used in computerised body scanners, sometimes called CT body scanners. It consists in measuring the strength of X-rays passed through the body of the patient from a number of directions, and using these measurements to build a three-dimensional image displayed on a screen.

computerphobia A fear of or aversion to computers.

Computing Services Association (CSA) A UK trade association of companies that offer services in the field of computing. Membership includes computer bureaux, facilities management companies, software and systems houses, and consulting firms.

concentrator A piece of equipment used within a data communications network to concentrate the traffic from a number of transmission lines on to a single higher speed line, and vice versa. A concentrator differs from a multiplexer, which does a similar job, in that it stores incoming messages before interleaving them in a single stream on the higher speed line, while the multiplexer divides up the capacity of the higher speed line into a number of channels and directs the traffic from each of the slower lines into one of those channels.

conceptual schema A schema describing data in conceptual terms, independently of the

55

means used to store it (represented in the internal schema or schemas) and the particular applications that will use it (the external schemas). In database terminology, a schema is a programmed description of data items (sometimes described as entities) and of the relationships between them. The conceptual schema is used to keep track of all the data used by a group of applications affecting a particular area of operations.

concurrency A measure of the capacity of a teleprocessing system. It defines the number of transactions that may be active, or in other words in the course of processing, within the system at any particular time.

concurrent DOS (CDOS) A personal computer operating system developed by Digital Research Inc. It is a multi-user operating system (whereas most personal computers run single-user operating systems at present) capable of running programs written for PC-DOS/MS-DOS (the current market leader) and CP/M (the former market leader) at once.

conditional branch A branch that depends on the value of a specified variable. The variable will have been set as the result either of a preceding comparison (such as of two numbers) or a preceding arithmetic operation (such as if the result is negative).

conditioned line A telephone line with additional equipment installed to make it suitable for data transmission.

confidence factor The degree of confidence that can be placed in a conclusion contained in a rule in a knowledge base. This can vary from +100% to -100%, +100% meaning absolute confidence that it is true, and -100% absolute confidence that it is false.

configuration The precise make-up of a computer system, expressed in terms of the various units of hardware and software that it comprises.

connection-oriented network service (CONS) Used of data communications services where devices wishing to exchange messages must establish a connection before sending one another messages, much as with a telephone call.
 See also *call establishment*.
 Compare *connectionless network service*.

connectionless network service (CLNS) Used of data communications services where the device wishing to send a message does not need to establish a connection with the receiving device before sending it. Messages sent using connectionless services are known as 'datagrams', by analogy with telegrams.
 Compare *connection-oriented network service*.

connectivity A measure of the switching capability of a telecommunications network or system, or in other words how easy it is for users of that system or network to establish connections with other users.

connector See *logical operator*.

CONS Abbreviation of *connection-oriented network service*.

consequent-driven reasoning See *backward chaining*.

consequent rule The type of rule that is applied in a backward-chaining expert system. It takes the general form 'IF A and B are true, THEN C must also be true'.
    Compare *antecedent rule*.

console log The continuous listing produced on the console typewriter of a computer system. It records all the messages entered by computer operators and those generated by the operating system.

console typewriter See *operator console*.

constant A data element within a program whose value is assigned by the programmer before compilation and does not change during execution of the program.
    Compare *variable*.

constant angular velocity (CAV) A method used to address information stored on magnetic storage media such as disk. Space on the disk is formatted into a series of concentric rings called tracks, and these in turn are divided into sectors. The disk drive locates any particular sector by positioning the read/write heads over the track and waiting until the sector appears underneath them.
    Compare *constant linear velocity*.

constant linear velocity (CLV) The method used to address information stored on compact disk. Information is stored in a spiral, as on gramophone records, and to find an individual item the disk drive has to read the whole spiral. This makes access much slower than with the track and sector addressing used on hard and floppy disks, known as constant angular velocity.

construct An arrangement of program statements that will produce a given effect when executed. High-level languages, for example, use constructs known as control structures, and most programming languages permit loops and subroutines.

content-addressable memory See *associative retrieval*.

contention Where two or more devices compete for the use of a transmission channel. The devices will normally follow a particular protocol (formally specified procedures) which resolves conflicts and ensures that the transmission channel is used efficiently. Data terminals using public telephone networks, for example, use a simple contention protocol (often called teletype protocol) which ensures that the line is used alternately. Many local area networks use a more sophisticated protocol known as carrier sense multiple access (CSMA), which ensures that collisions – attempts by more than one device to send at the same time – are detected and that the devices re-transmit as soon as possible.

Compare *token passing*.

context A concept used to group ideas, objects, or any other items that certain types of expert system work with.

context-sensitive Varies depending on the circumstances in which called on. Used particularly of help services on terminals or personal computers where the response reflects the task that the user is engaged in at the time.

context-switching A term used to refer to difficulties people experience when switching between applications on a computer system. This arises because they need to adjust to changes in the context in which successive applications operate, such as the varying methods used to present information and to control the application.

context tree A diagram showing how the contexts of an expert system are related.

continuous stationery Special stationery used by computer printers, consisting of a continuous length of paper with regular perforations so that it can be separated into pages after printing. Also known as *fanfold stationery*, because it is normally folded like a fan at each perforation.

control character A character used to control the operation of a transmission line or of a device such as a printer or a visual display terminal. For example, a paper feed control character specifies how many blank lines are to be skipped before the next line is printed; communications control characters are used to tell terminals when to start or stop transmitting, and to mark the beginning and end of text in a message to be transmitted; cursor control characters are used to position the cursor on a visual display screen. Contrast with a *printable character*, that can be printed out on paper or displayed on a screen.

control flowchart See *state transition diagram*.

control register See *address buffer*.

control structure A means of controlling the sequence in which the statements of a program are executed. Assembly languages use two simple control structures – branches and loops. Modern high-level languages such as Pascal or PL/1 include a range of powerful control structures such as *IF*, *CASE*, *REPEAT*, *FOR* and *DO WHILE* statements.

control total A total accumulated by adding together one or more of the fields in a series of records or documents involved in a series of processing operations. These totals are checked at each stage of processing to establish that all records have been processed, without needing to check each individual item.

control unit The device that controls and co-ordinates the operation of a computer processor. One of its main tasks is to determine the sequence in which program instructions are to be executed, and to fetch and decode those instructions prior to execution.

In more detail, the control unit operates as follows. For each program instruction, it operates in two cycles. It begins by fetching the instruction from computer memory, then decides what operation is to be performed, identifies the operands and, if necessary, fetches those also from memory. This is known as the fetch cycle. It then enters the execute cycle, activating the arithmetic/logic unit to carry out the processing, then returning the results to memory. Together, the fetch and execute cycles are known as the instruction cycle, and generally the instruction cycle takes between one and three processor cycles to complete.

But the control unit is not only concerned with executing program instructions. Some of those instructions (READ instructions and WRITE instructions) initiate transfers into and out of the memory from peripherals such as disk drives and printers. After initiating these transfers, the control unit must monitor their progress, then start up a special sequence of instructions when the transfer is complete (see *interrupt handling*). Because these transfers are very lengthy in comparison with the instruction cycle, the control unit continues to execute program instructions while transfers are in progress. To improve performance, the control units of advanced computers also use a technique known as pipelining to overlap the execution of the various cycles of different instructions.

control variable A variable which is used to control the number of times that a loop is executed. It is initialised to a required value, then increased or decreased each time the loop is executed until a specified value is reached.

convergence A measure of the quality of a colour display, reflecting how precisely the beams from the red, green and blue electron guns in the display hit their target on the screen. This is evident, for example, in the colouring of straight lines.

See also *colour purity; colour uniformity; grey linearity*.

conversational See *interactive mode*.

conversion See *code conversion* or *protocol conversion*.

cooperating sequential processes Processes within a computer system that run independently but must coordinate their activities, for example because they both update the same data. Such processes are found within transaction processing and distributed processing systems and are potentially the source of program errors. In the 1970s a Dutch computer scientist called Edsger Dijkstra first proposed a scheme for controlling them.

coprocessor A processor chip that works in conjunction with the main processor in a computer system, to perform particular operations more efficiently. Some personal computers, for example, have a maths coprocessor chip to speed up floating point calculations.

copy and paste See *cut and paste*.

copy protection Techniques, usually applied to floppy disks containing proprietary applications packages, that make them impossible to copy using the normal utilities. This is done to prevent copies being made and distributed in violation of restrictions imposed by the owner of the copyright in the software. Copy protection techniques usually involve formatting the floppy disk in a special way that allows it to be read but makes it impossible to copy correctly or completely. Alternatively, a master disk is supplied with a special imprint that cannot be copied. Copies can be made of the master disk, but applications run from these copies call for the original master disk when started up and will not run until they have verified that the correct imprint is present.

See also *intellectual property*.

CORAL A high-level programming language designed for real-time applications.

core storage Computer memory before 1970, consisting of ferrite cores – tiny rings of magnetic material 1mm or so in diameter – strung by the thousand on grids of wires. Now superseded by semiconductor memories – see *random access memory*.

corporate data model A data model representing the business entities belonging to a business or an organisational unit, and showing the relationships between them. This is a high level model expressed in conceptual terms, rather than in terms of how information will be processed or stored by any computer-based systems. An example applying to a hospital is shown opposite.

A corporate data model will normally be drawn up by means of an intensive systems planning study and is referred to by a number of alternative terms including enterprise model, strategic data model and data architecture.

corrective maintenance Maintenance of computer equipment undertaken after equipment fails.

Compare *preventive maintenance*.

corrupt Accidentally overwrite or alter in such a way that the information affected can

Example corporate data model

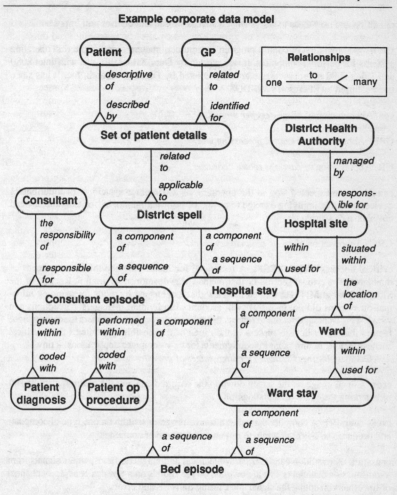

no longer be relied on. Used particularly of information that is sensitive to even minor changes, such as a program in memory or a directory on a disk. For example, information on disk can be corrupted if the read/write heads on the drive go out of alignment.

cost-based optimiser A feature of relational database management systems, designed to optimise efficiency of processing. As the system runs it gathers statistics about its

61

operation and uses these to assess what demands subsequent queries will impose.

CP/M Abbreviation of control program for microcomputers – one of the first operating systems for personal computers. It was written by Gary Kildare for use with Intel 8080 and Zilog Z-80 microprocessors and is marketed by Digital Research, Inc. It has since been eclipsed by Microsoft's MS-DOS.

cps Abbreviation of *characters per second.*

CPU Abbreviation of *central processing unit.*

CR Abbreviation of *carriage return character.*

crash An unexpected stop in the operation of a computer system or of a peripheral device, probably implying damage to or loss of information. As in, for example, a 'head crash' on a disk drive.

CRC Abbreviation of *cyclic redundancy check.*

critical success factors (CSF) A technique for identifying the information needs of managers. The critical success factors method was developed by John F. Rockart of the Sloan School at MIT. He formed the view that established methods for planning information systems did not take sufficient account of individual managers' perceptions and views. This technique set out to redress the balance. It requires managers to identify those factors that are critical to success in their area of responsibility. Some of these 'critical success factors' in turn point to requirements for new computer applications.
Compare *business systems planning*; *stages of growth.*

crop Cut an image on the screen down to the required content; the electronic equivalent of trimming the edges off a photograph.

cross compiler A compiler that takes a source program written on one type of computer and produces an object program to run on another type of computer.

crosstalk A condition caused by stray electrical and magnetic fields, where signals from a transmission channel or line are overlaid on the signals on a physically adjacent channel or line, thus corrupting the signal and causing transmission errors.

CRT Abbreviation of *cathode ray tube.*

crystal bistability See *optical bistability.*

CSA Abbreviation of *Computing Services Association.*

CSF Abbreviation of *critical success factors.*

CSMA Abbreviation of *carrier sense multiple access.*

CST Abbreviation of *computer-supported telephony.*

CT Abbreviation of *computerised tomography.*

CTS Abbreviation of *clear to send.*

CUG Abbreviation of *closed user group.*

cursor The visual indicator on a display screen, showing where the next character typed on the keyboard will appear. Sometimes a flashing square block is used, sometimes an underline, sometimes a vertical bar.

cursor control The positioning of the cursor on a visual display screen so that information is displayed and/or to help users to enter data in the right places. Applications programs control the cursor by sending special control characters to the visual display unit, or by including these or escape sequences in the messages sent to a visual display terminal.

cut and paste A method of transferring information from one place to another or from one application to another, on a personal computer. First the information to be transferred is 'cut' from its source, which has the effect of placing it in a temporary storage area. Then the user moves on to the destination – a new application or a new position in a document – and 'pastes' into the selected position, to where the information is transferred from the temporary storage area.

cut sheet feeder See *sheet feeder.*

cutover The time when the procedures that a new computer-based system requires are adopted, replacing those in use prior to its introduction.

cybernetics The study of systems of control and communication, both in animals as well as in electronically-based systems. The father of cybernetics was the mathematician Norbert Wiener whose book by that title was published in 1948. One of the key aspects of cybernetics is the concept of feedback – the ability of a system to use the results of its own performance as information to regulate its own behaviour. If this feedback information is used wisely, then the system remains stable as external conditions change. If not, then it enters into a cycle either of decline or of steadily increasing activity.

cyberphobia See *computerphobia.*

cycle time In general, the time taken to complete a cycle of operations. Computers are often characterised by the cycle time of the processor, which is the time the control unit needs to complete any single operation. This varies from around 10 nanoseconds in so-called supercomputers, down to about a microsecond in a low cost personal computer (although in a typical business PC it is about a tenth of that). This is often expressed as a frequency in millions of cycles per second (or MHz). Hence processors vary in speed between 1 and 100 MHz. Processor cycle time does not, however, equate directly with instruction throughput – see *execution time*.

Computers are also characterised by their memory cycle time – the minimum time needed to retrieve a word of data from memory. Because costs of very high speed memory would be prohibitive, this is always considerably slower than processor cycle time in large-scale computers, although some personal computers with much slower processors have memory of equal speed.

cyclic redundancy check (CRC) A method used to check the accuracy of digital information sent over a communications link. A complicated formula is used to calculate a field, typically consisting of 8 or 16 bits and sometimes called a frame check sequence, from the bits contained in the information field of a message. This is appended to the message before it is sent. The receiving device does the same calculation, and compares the result with the check field it has received. If the two do not match, it asks the sending device to retransmit.

cylinder On a disk pack, all the tracks in one circular slice down through all the surfaces, or in other words all the tracks that are immediately under a read/write head when the arm is in a particular position. Normally, a disk drive has one read/write head for each surface, mounted on an arm which moves them all in and out over the surfaces together.

D

D/A Abbreviation of *digital-to-analog conversion*.

daisy chain A method of attaching a number of peripherals to a single interface on a computer or to a single controller. A connecting cable is carried from the interface or controller to one of the peripherals; a separate cable connects the first to the second; then the second to the third, and so on. The final device in the chain has a blanking plug in its onward socket, signifying the end of the chain.

daisywheel printer A type of printer widely used in offices to print letters and reports, producing results of similar quality to those produced by a typewriter. Daisywheel printers have a rotating metal or plastic wheel, each spoke of which has the raised shape of a character. For each character to be printed, one of these is brought under the print hammer and struck against the ribbon and, in turn, the paper. Daisy wheels can be replaced to change font or typestyle, and printers usually have internal switches that can be set for different sizes of paper, different line spacing, and so on. These values can also be set automatically by the applications program in the computer to which the printer is attached.

DAL Abbreviation of *data access language*.

darkroom computing Operation of a computer room without human operators. Operating systems for larger computers have been evolving in this direction for some time, with the aim of reducing operating costs.

DASD Abbreviation for direct access storage device, usually pronounced 'dazzdy'. See *direct access storage*.

DAT Abbreviation of *digital audio tape*.

data A general term used to describe the raw material that is processed by a computer system, or indeed by any other means. It is used as a collective noun usually accompanied by a singular verb – 'the data is processed ... '. ('The data are processed' is semantically correct, but is awkward to say and is rarely used.)

The term is used more precisely to draw a number of distinctions:

(1) Data is raw facts as opposed to meaningful *information*. In other words, data must pass through a computer program or some other preparatory process to be decoded and put into context before it becomes information that makes sense to a human being. Thus 'A12345' is data that could mean almost anything; 'product code A12345' might make sense to a stores clerk; and 'product A12345: can of baked beans' makes sense to most people.

(2) Data is information in a highly structured form, consisting of codes, numbers and structured text such as names and addresses or product descriptions, as opposed to information in other immediately accessible forms – *text*, as in documents; *image* or *video*, that can be looked at; *voice*, that can be listened to.

(3) Data is the information held within a computer system that is acted on, as opposed to the *program* that decides what action should be taken with that data.

data access language (DAL) A language, designed by Apple Computer and licensed to a number of other major computer manufacturers, intended as a standard means for applications and database packages running on different computer systems to exchange data. It is an extension of structured query language (SQL), allowing any type of data to be exchanged, not just data from relational databases. Unlike SQL, which is solely a language definition, each system that supports DAL has a special driver program that acts as intermediary with programs using the language.

See also *structured query language*.

data administration A role associated with data management and the concept of 'data as a resource', arising from database management. Data administration deals with the business and political issues arising from the sharing of data across organisational units. Contrast with *database administration*, a lower level role concerned with the technical aspects of managing a shared database.

data alignment How data structures are aligned (for example with respect to word boundaries) in computer memory. Data alignment may become an issue when programs are moved ('ported') from one type of computer to another.

See also *byte ordering*.

data analysis Examination of an organisation or some aspect of its activities in terms of the data it requires to operate. Contrast with traditional *systems analysis*, which concentrates on the procedures an organisation carries out. A data analysis exercise begins by identifying the principal entities on which the organisation depends – customers, suppliers, products, branches, accounts, orders, invoices, etc. It then defines their attributes and relationships with one another. Finally, the entities and their interrelationships are described in what is known as a data model.

data architecture See *corporate data model*.

data bank A general term for a large amount of data stored on computer files, usually on a specific topic. Normally it would also be accessible from remote terminals, such as via an information retrieval service.

data broadcasting One-way distribution of data via television sets. As for teletext (such as Oracle and Ceefax), broadcast data is injected into the video stream just before transmission, using the spare capacity left over from the TV signals. It can only be received by a user equipped with a special decoder which selects only the information addressed to that particular type of decoder. This means that data can be addressed to individual users anywhere within the reach of television broadcasts. It can also be encrypted for security.

data bus See *bus*.

data capture A general term meaning the entry of data into a computer system, via a keyboard or by some other means, so that it can subsequently be processed. In the form 'source data capture' or 'data capture at source', it means capture of data at the point where it originates, such as in a bank branch or at a supermarket checkout. The advantage of source data capture, using online terminals, is that data can be validated and corrected on the spot before it enters the processing cycle.

data centre A location containing shared computers and the staff that operate and (sometimes) program them. Also known as *computer centre*.

data circuit terminating equipment (DCE) A general term used by telecommunications authorities to mean the equipment which connects a user device such as a terminal or computer (known as the data terminal equipment or DTE) to a transmission line or public network. On public telephone networks the DCE is a modem, but on digital data networks what is called a network interface unit is used instead.

data collection The capture of data at a number of points at which it originates, from where it is transferred electronically to a central point at which it is to be processed.

data communication A general term for the transfer of coded information, and particularly numbers and text, within and between computer-based systems. The key feature distinguishing data communication from telephony, or in other words voice communication, is that communication need not take place in real time. Depending on the particular application, delays between the sending of a message and its receipt are acceptable varying from milliseconds to minutes or hours.

See also *mobile data communications.*

data compression Reduction of the number of bits used to represent information, thus limiting the amount of storage space or transmission capacity needed to handle it. The key to compression techniques is the replacement of repetitive patterns in the data by shorter codes representing each pattern. A number of different algorithms are used to achieve this, and these can be divided into two categories.

'Lossless' algorithms are used for information such as programs or data processing files where no loss of detail is acceptable. For example, a technique called run length encoding measures and records the length of continuous sequences of 0s or 1s, rather than recording the sequence itself.

'Lossy' algorithms (such as the JPEG algorithm) are used for information such as colour images where some detail can be lost with no significant loss in the quality of the final image, achieving higher compression ratios.

See also *JPEG algorithm*; *LZW algorithm*; *Huffmann encoding*; *MPEG algorithm.*

data declaration A program statement that tells the compiler about an item of data. Data declarations are used to specify the variables, constants and data structures to be used by the program.

data description language (DDL) A programming language used to describe the structure and content of data files and the relationships between them (often referred to as schemas). A data description language is included as one component of many database management systems. Contrast with *data manipulation language* – used to write the applications that use the database files.

data dictionary A utility that stores details of the attributes of data items used by applications programs. This 'data about data' is sometimes referred to as metadata. The data dictionary records metadata in a disciplined way intended to prevent incomplete or inconsistent definitions of data. The details stored for each data item normally include an explanation of the meaning of the data; the length and format of the field, such as whether it is an integer value or alphanumeric; the range of values it is permitted to have; and which applications programs use it. The dictionary also records the relationships of data items to one another. Some also define other aspects of the way applications programs operate, such as screen formats, the contents of menus and other standardised procedures. These extended data dictionaries are sometimes referred to as encyclopaedias or repositories.

A data dictionary may be passive – in other words kept up-to-date by separate clerical procedures – or active – automatically kept up-to-date by the database management system whose files it describes.

data division The part of a COBOL program that associates each data file that is to be processed with a peripheral, and shows the layout of the records held in the files and of any work areas and constants that the program will use.

data-driven reasoning See *forward chaining.*

data encryption standard (DES) A standard specifying procedures for encryption and decryption of data for secure transfer over a communications link. It was developed originally by IBM and, after review, affirmed by the US National Institute for Standards and Technology as an official standard. DES is widely used for in the US for funds transfer and other applications. Software incorporating the standard can only be transported overseas with US Government approval. The technique is regarded as unbreakable, but it depends on a bilateral agreement between sender and recipient on the key to be used, which must be changed regularly to maintain security. With the other widely-used encryption method, RSA, management of encryption keys is simpler.

data entry The process of entering data manually into a computer system, usually via a keyboard, following which it is stored on mass storage, such as disk or tape, ready for subsequent processing. Data entry devices have been widely used in the past to capture data prior to processing in batch, but this method is being superceded by data capture at source using online terminals and by electronic methods of capture such as optical character recognition (OCR).

data field A meaningful unit of information, and hence the unit in which data is normally entered via a keyboard and/or display screen, retrieved via a query language, or manipulated by an applications program. A data field is normally referred to by a name. Records are made up of one or more data fields, each field in the record representing one of the attributes of the entity represented by that record. An employee record within a personnel system, for example, might have fields for 'age', 'name', 'address', 'current salary', 'date of employment', and so on. Programs also use data fields internally to hold temporary results during processing.

Each data field will be of a particular data type, such as a string (of characters), an integer, or a real number.

data file An organised collection of records, serving as a basic unit of storage on mass storage devices such as disks or tapes, sometimes known as a data set. Records in a data file have a consistent format throughout the file. They may be related in terms of their source or destination (for example, a file of transaction records originating from a particular office, or a file of information to be printed); the entity they represent (such as a product file or a customer file); or their purpose (for example a journal file used to record all activity in a transaction processing system).

data flow diagram A diagramming technique used by systems analysts. A data flow diagram maps the flow of information through a business or through a computer-based system. It shows external agents (such as customers), actions by people (such as

69

accepting an order or despatching goods), and actions by the computer system (such as processing invoices or storing a record in a data file). Arrows between these represent the flow of information and goods. (The figure shows the main symbols used.)

Compare *state transition diagram*; *action diagram*.

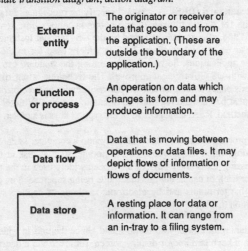

| External entity | The originator or receiver of data that goes to and from the application. (These are outside the boundary of the application.) |

| Function or process | An operation on data which changes its form and may produce information. |

| Data flow | Data that is moving between operations or data files. It may depict flows of information or flows of documents. |

| Data store | A resting place for data or information. It can range from an in-tray to a filing system. |

data haven A country with loose data protection laws.

data independence The concept that data should be independent from the programs that use it and from how it is stored, so that either can be changed without affecting the other, and so that people (programmers, end-users) using the data do not need to know things that are irrelevant to their particular tasks. Data independence can take a number of different precise forms, and can be separated into two broad categories:

(1) Physical data independence is where the data is independent of the methods and devices used to store it, so that, for example, the physical organisation of files can be changed without affecting programs.

(2) Logical data independence is where the overall view of the data (the conceptual schema) is separate from that of individual users and programmers (external schemas), so that they do not need to know about the overall structure of the database and so that the database can be extended without affecting existing users and programs.

data interchange format (DIF) A format for the exchange of data files between personal computer applications.

data item The smallest useful unit of information processed by an applications program. This may be a data field, or a subdivision of one. For example, the data field 'date of

70

employment' may be treated as a single item or as consisting of three items for day, month and year respectively.

See also *elementary item*.

data link layer The second protocol layer of the reference model for open systems interconnection (OSI) defined by the International Standards Organisation (ISO). The data link layer provides a reliable link between two adjacent nodes or user devices in a network.

See also *application layer*; *network layer*; *physical layer*; *presentation layer*; *session layer*; *transport layer*.

data link level See *data link layer*.

data management The formal collection, storage and preparation of data, to ensure that it is available and appropriate for the uses to which it may be applied within an organisation. Data management uses techniques such as data analysis, and software tools such as data dictionaries and database management systems. It frequently, but not necessarily, leads to the development of computer-based database files. The administrative tasks associated with data management are referred to as data administration.

data manager See *database manager*.

data manipulation language (DML) A programming language used to write the applications that use the data files within a database. Contrast with *data description language*, which describes the structure and content of data files and the relationships between them.

data mining Deriving useful information from large databases, using techniques such as statistical and sensitivity analysis.

Data mining can be thought of as the reverse of an expert system. Where an expert system is given the rules and then uses them to draw inferences based on given facts, data mining attempts to discover rules hidden in a set of facts recorded in a database. For example, after a lead poisoning survey threw up baffling results, data mining software was used to trawl the survey database and revealed patterns based on gender which should not in theory have affected the outcome.

data mode The state of a modem when it is receptive to data signals. Contrast with *talk mode*, when signals are allowed straight through the modem to the telephone handset.

data model A representation of the data affecting a particular area of an organisation's operations. It normally consists of diagrams and text describing relevant objects, events and concepts, together with their main attributes, the relationships between them, and a set of rules defining how they may be manipulated. A corporate data model is used as a framework for an organisation's information systems overall. Entity-relationship models are used as a guide to the subsequent design of applications programs, and data models based on hierarchical, network and relational models are used as the basis for the design of particular databases.

data over voice (DOV) Technology that enables voice (i.e. telephone) and data (i.e. computer) users to share wiring. Data is sent over a different frequency spectrum than voice and is separated by equipment installed at each end of a connection. Neither telephone nor computer users are aware that they are sharing wires.

data path The number of bits that can be transferred along the bus between memory and processor at one go, and hence a measure of the speed at which a computer will operate.

data preparation See *data entry*.

data processing (DP) A general term meaning the application of computers to administrative processes such as accounting and stock control, or in other words most of the uses to which computers were put initially in the commercial area. With the widening application of computers, it is now used to describe the application of computers to the basic operational systems of organisations, as opposed to their application elsewhere, such as to support managers or professional staff or in the factory.
   Compare *information processing*.

data processing department See *information systems department*.

data protection Legislative measures designed to prevent the misuse of personal data held on computer systems. Most Western countries have legislated to require organisations holding personal data about individuals to register details. They are usually required to provide mechanisms for individuals to inspect data relating to themselves and to have it corrected if inaccurate. In the UK, this legislation is embodied in the Data Protection Act and administered by the Data Protection Registrar.
   See also *data haven*.

Data Protection Registrar An official appointed by the UK Government to ensure that the provisions of the Data Protection Act are complied with. The Registrar keeps a register of organisations whose computer systems fall within the scope of the Act – because they hold certain types of personal data on individuals – and also investigates complaints by individuals that data relating to them is inaccurate or is being abused.

data security officer A person who has responsibility for making sure that data stored on computer files remains secure against unauthorised access, accident or sabotage. The role may also cover data stored locally, such as on disks attached to personal computers.

data service unit See *limited distance modem.*

data set (1) In the telecommunications world, a device which performs any control and conversion (such as modulation/demodulation) functions necessary to match a device such as a computer or terminal with a telecommunications service. Examples are modems and line drivers.

(2) In the data processing world, a data file.

data storage The use of any device to store data in a permanent or semi-permanent form, such as on disk or magnetic tape.

data structure A logical framework for a number of related data fields which enables the programmer to deal with them either as a single unit or individually. Examples include arrays and records.

data terminal (1) Used within telecommunications to mean any device capable of sending or receiving digital information over a communications network – see *data terminal equipment.*

(2) Used within computing to mean a device that operates under the control of a computer system and to which it is connected by means of a communications link.

data terminal equipment (DTE) A general term used by telecommunications authorities to mean any user device, such as a terminal or a computer, that is connected to a data transmission line or to a public network. The data terminal equipment (or DTE) is connected to the transmission line or to the network by data circuit terminating equipment (the DCE) such as a modem or a network interface unit.

data transfer The process of transferring data (meaning digitally encoded or non-voice information) from one point to another. Also used to distinguish the central phase of a non-voice call from those which precede and follow it, namely call establishment and call clearing.

data transfer rate The speed at which data bits are transferred along a transmission channel, normally measured in bits (or kilobits or megabits) per second.

data transmission The process of transmitting information in digital form, in other words as coded electrical pulses, or the techniques associated with that process.

data type The way a data field (and the bit pattern that represents it in memory) is to be interpreted. In the BASIC language, for example, data fields may be defined as one of two main data types: real numbers or strings of characters. Depending on which is specified, the compiler stores and processes the field in the appropriate way.

See also *abstract data type.*

database (1) In technical terms, a structured store of data; that is, a store containing both data and the means of maintaining the relationships between the data. These relationships in turn reflect the relationships between the real entities – physical objects, events and abstract concepts – described by the data.

(2) In application terms, a data file or a set of data files designed to reflect the nature of the data they hold, rather than the needs of particular applications that process them. This has the great advantage that a set of integrated files can be created to serve a number of related applications (see figure). This reduces inconsistencies and eliminates the need to transfer information between them.

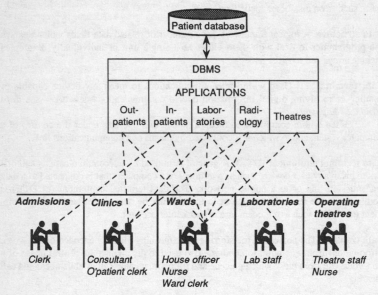

(3) The term is often used more loosely, to mean any collection of related data files.
See also *database management system*.
Compare *knowledge base*; *flat file*.
More detail *network model*; *relational model*.

database administration (DBA) A role associated with database management and the sharing of data between different applications and users. Database administration is concerned with the technical aspects of managing a database, such as maintaining the schemas which give different users their particular 'view' of the shared data. Contrast with *data administration*, a higher level role which deals with the business issues arising from the sharing of data across organisational units.

database engine See *database machine*.

database integrity A highly desirable condition for a database, in which the content of records in the database and the relationships between records are accurate and consistent. Integrity is a particular problem in database files both because the files may be shared between a number of applications and users, and because the file structures may be complex. This makes it very difficult to reconstruct the precise series of events leading up to a failure of the computer system maintaining the database, so that the files can be restored to a known state ready for resumption of processing. Some database management systems allow database designers to define the integrity conditions to be applied to particular records. These are either checked automatically whenever such records are updated or can be checked periodically by running a special utility program.

database machine A processor which forms part of a larger computer system and whose function is to manage a database at the dictates of the computer or computers running the applications programs. Normally, the database machine will be installed at the same location as the controlling computer(s), with a high-speed interconnecting link. A database machine has several advantages compared with running the database management software in the same processor as the applications programs: (1) the same database can be accessed by two different types of computer if need be and (2) it can be expressly designed for the task of searching and updating database files, rather than to handle a general computing workload. It may, for example, use associative retrieval techniques, rather than the one-record-at-a-time searches used by conventional computer systems.

database management system (DBMS) A set of interrelated software tools designed to construct and provide access to a database held on a computer system, and to control the privacy, security and integrity of the data. The tools usually embody special programming languages for describing and manipulating the data, known as data description and data manipulation languages respectively. They may also include a data dictionary system – a means of recording the contents of a database and how it is used.
See also *database administration*.
Compare *file management*.

database manager The software component at the heart of a database management system. The database manager records changes to the database files, translating the logical access requests issued by applications programs into terms of physical storage, and ensures that the integrity and security of the data is maintained. Note that the people closely involved with the management of databases are referred to as *data* or *database administrators*.

database update See *record update*.

dataflow computer An innovative form of computer in which the sequence of operations is determined by the flow of data values, rather than by the logic of a program as in the conventional *von Neumann* architecture of most of the computers in use today.

datagram A message sent over a data communications network using a connectionless protocol. In other words, the device wishing to send the message does not need to establish a connection with the receiving device before sending it, but merely posts it into the network like a telegram.

See also *connectionless network service*.

daughterboard See *expansion card*.

DBA Abbreviation of *database administration*.

DBF The proprietary format defined by Ashton-Tate for database files and incorporated into the company's highly successful personal computer package, dBase. A number of other database packages are able to import files in this format.

DBMS Abbreviation of *database management system*.

DBS Abbreviation of *direct broadcasting by satellite*.

DCA Abbreviation of *document content architecture*.

DCA-RFT Standing for Document Content Architecture/Revisable Form Text, this is a widely-used format for the exchange of word processing documents, originally defined by IBM. Many popular word processing packages accept and allow users to save documents in this format. It preserves not only the text content of the original document but also its format, although not necessarily in every detail.

DCE Abbreviation of *data circuit terminating equipment*.

DDI Abbreviation of *distributed data interface*.

DDL Abbreviation of *data description language*.

DDP Abbreviation of *distributed data processing*.

deadly embrace A deadlock condition that can arise where two or more applications programs (or threads – see *multithreading*), running concurrently, are accessing shared files. The condition arises when each of two such programs can only proceed when it can gain access to a record that the other is holding. For example, each program wishes to retrieve two records from the shared files before updating both of them. Program 1 retrieves and locks (i.e. reserves for exclusive use) record A, then attempts to retrieve record B. Meanwhile, program 2 has retrieved and locked record B, and is now attempting to retrieve record A. The solution is usually imposed by the operating software, which times how long programs are suspended waiting for records. If a program exceeds a specified threshold time, the operating system forces it to release all its records and restart. This allows the other program to complete and breaks the deadly embrace.

Debit/Credit A benchmark for transaction processing systems, developed by Bank of America in 1973. It uses one simple transaction consisting of 25 COBOL statements, and indicates the throughput obtained when 95% of transactions are completed in one second. Debit/Credit figures are quoted by a number of major computer suppliers, including DEC and ICL. There are several variants of Debit/Credit, and notably *ET1* and *TP1*.

debug Remove the errors – the bugs – from a program, by checking it and running tests.

debugger See *interactive debugger*.

decision analysis Methods for analysing preferences prior to making a decision, including but not limited to those supported by computer.

decision model See *decision tree*.

decision support system (DSS) A computer system designed to help people to make decisions. It will normally provide some means of capturing and storing the data on which decisions depend, plus various tools that can be used to manipulate the data, in order to model alternatives and explore the consequences of different courses of action. Although originally envisaged as applicable to a whole range of management decisions, decision support systems have proved effective mainly in situations involving complex but predictable scheduling algorithms, such as train dispatching or air traffic control.
 Compare *executive information system*.

decision table A way of showing the basis on which decisions are made, by means of a table showing the relationship between the variables involved and the actions to be taken under different conditions. As the figure shows, each table consists of four parts:

(1) the condition stub listing the conditions;
(2) the condition entry showing the permitted combinations of values for those conditions;
(3) the action stub listing the actions; and
(4) the action entry showing what actions are performed for the entries in the condition entry.

	Rule 1	Rule 2	Rule 3	...
Condition stub	Condition entry			
Action stub	Action entry			

 Decision tables may be included in system or program documentation, either to supplement or replace narrative and flowcharts, and can be processed by software to generate program statements automatically.

77

Compare *decision tree*.

decision tree A way of representing a decision-making process which shows it as a series of questions linked in a tree structure. Decision-makers (or computer programs that model the process) start at the base of the tree and the answer to each question sends them along a particular branch until they arrive at the final decision, at the outer edge of the tree – see simple example below.

Compare *decision table*.

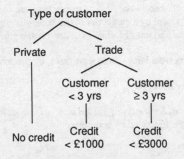

declarative graphics See *object graphics*.

declarative language A programming language in which the programmer declares certain facts about a situation, leaving the language compiler to draw appropriate conclusions from those facts.

Compare *procedural programming*.

declare Specify the identifiers to be used in a program, so that the compiler can process them in the appropriate way and, for data fields, reserve space in memory.

See also *data declaration*; *data division*.

DECnet The trade name for Digital Equipment Company's set of products for the interconnection of its computer systems and terminals within a network. Digital is the second largest computer supplier in the world after IBM and the largest single supplier of minicomputers.

Compare *open systems interconnection*; *systems network architecture*.

decoder The component within an adapted television set that decodes incoming teletext (broadcast along with the normal video signal) or videotex (received over a telephone line) messages, and displays the frames on the television screen.

decoding/control unit See *control unit*.

decryption See *encryption*.

dedicated line See *leased line*.

deductive database A database that holds rules as well as data and that applies these rules automatically when the database is altered. A conventional database, by contrast, holds only data, and all rules and algorithms applied to alter the data are embodied in applications programs.

deep knowledge General, theoretical knowledge, in other words knowledge that is anchored in first principles, axioms and laws. This is in contrast with *shallow knowledge* or expertise, developed through training and experience. The terms are used in the expert systems field.

default A preset response to a question or choice from a menu presented to a user by a computer system. The default represents a common or frequently held value. It is normally selected either by taking no action at all or by a simple action such as pressing the RETURN key, or where appropriate can be overridden. For applications requiring a large amount of user input, defaults can save the user both time and annoyance.

delete (1) Of data files and records, eliminate by complete removal from the storage device.
(2) Of programs running in a computer system, remove from memory.

delimiter A symbol (or a pair of symbols) used to indicate where one field in a record ends and the next one begins. For example, in ASCII text files comma or TAB characters are used as delimiters to separate fields within each record and a carriage return character is used at the end of the record, while in Pascal programs single quotes are used to delimit strings and curly brackets or * are used to delimit comments.

delta YUV (DYUV) A method for encoding signals representing video images, to reduce the space needed to record and store them electronically. This technique is used in compact disk interactive (CD-I) technology, allowing over 6,000 colour images to be stored on a single compact disk.

DENDRAL An expert system that helps organic chemists to identify organic compounds by analysing mass spectograms. It was one of the first expert systems to be developed, work beginning at Stanford University in 1965.

departmental computing A form of computing in which departments or groups in an organisation control, and perhaps operate, their own computer systems, which process a range of applications to satisfy departmental objectives. It might involve a dedicated minicomputer with terminals, or a number of personal computers linked together and to shared resources, such as departmental data files held on disk. Work group computing is almost synonymous, and perhaps a better term because it discards the connotation of organisational structure – departmental computing does not always respect existing organisational boundaries.

departmental processing See *departmental computing*.

DES Abbreviation of *data encryption standard*.

descender The 'tail' on a character such as 'g' or 'p' that descends below the line of text. Printers and display screens able to create true descenders produce text that is more readable than those that cannot do so.

desk checking Test a program without running it, by working out in detail what it would do given a range of possible inputs.

desktop The initial display on a personal computer with a graphical user interface, such as the Apple Macintosh. By analogy with a real desktop, it shows various icons and windows overlapping one another, each representing applications, files and other features that are available.

desktop computer A personal computer designed to be installed on the desk, as opposed to smaller and lighter variants such as *laptop* or *notebook computers*.

desktop publishing (DTP) Publishing of, for example, newsletters, magazines, technical documents or sales material, using a computer system installed on the desktop. Typical components of a desktop publishing system are a personal computer with a high-resolution A4- or A3-sized screen; a laser printer or a photo-typesetting machine to produce the hard copy; applications packages for page layout, word processing and drawing illustrations; and perhaps a scanner to capture photographs or drawings. Desktop publishing offers total creative control of a publication from concept through to final output, and also great flexibility, since copy or design can be altered almost right up to the moment when the printed output is produced.

despatch Start up a task.
    See also *multitasking*.

destination When a file or document is copied, the duplicate as opposed to the original version (the *source*).

detachable keyboard A keyboard that can be detached from and moved independently of the unit containing the display screen and/or the processor. This makes it easier for users to find a comfortable working position. The keyboard is connected by means of a flexible cable or, occasionally, communicates by means of infrared signals.

development lifecycle The complete process of developing and then using a computer application. In the traditional approach to systems development, the lifecycle is seen as a linear process divided into a number of consecutive phases – beginning with a feasibility study or requirements study; followed by systems analysis, system design, programming, testing and installation; and concluding with an enhancement and maintenance phase (see figure).

One drawback of this linear lifecycle is that it is unnecessarily elaborate for simple applications, such as are implemented on personal computers. For such applications and when packages are used, some of these phases may be skipped or merged together. A

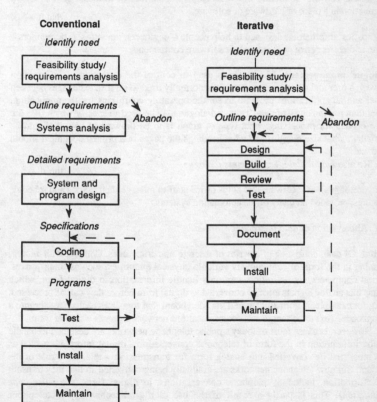

further drawback is that many applications exist in an environment of constant change, with the result that requirements change, sometimes drastically, between design and delivery. To address this problem, powerful system building tools are used so that the design-program-test phases of the lifecycle may be iterated rapidly. With each iteration, the design is progressively refined based on user experience (see figure).

See also *prototyping*; *system documentation*.

device driver See *driver*.

DIA Abbreviation of *document interchange architecture*.

diagnostic routine A routine which serves to trace errors or to locate the cause of a malfunction in a piece of hardware or software.

diagnostics Information designed to help people – engineers, operators, programmers – locate and correct errors in hardware or software components.

dialogue management The methods used to control the exchange of information between a user and a computer system. It is normally used to refer to exchanges between a user and an applications program to set the operating conditions for the application, rather than routine entry of data. Thus, a dialogue may be used to specify which files are to be opened for processing; what type of graph is to be drawn and how it should be formatted; what size margins and what headings the pages in a printed document should have; etc.

See also *modal dialogue*; *modeless dialogue*.

diary management Software that helps office staff to manage their diaries and schedule meetings, provided on many office automation systems.

DIF Abbreviation of *data interchange format*.

digital Of data, consisting of a series of discrete numeric values. Contrast with *analog*, meaning in the form of continuously variable physical quantities, such as sound waves. Digital computers, by definition, can only handle information in digital form, which means that analog signals must be converted to digital form before they can be processed. Text, for example, is normally entered via a keyboard that translates each character into a digital code. Text to be sent over a communications network assumes a similar form.

However, because most of today's public telephone networks are designed to handle analog information in the form of telephone conversations, digital information such as this must first be converted into analog form for transmission – this is the role of the modem. But now telephone networks are gradually being converted so that they transmit all information, including telephone conversations, in digital form (see *pulse code modulation*). This is partly a result of the use of digital technology in telephone exchanges, but also because analog transmission has a disadvantage for normal telephone traffic. Because the signal varies continuously and unpredictably, it is difficult to reconstruct it when it is distorted by interference. With digital signals, by contrast, consisting of a series of 0s or 1s, it is much easier to reconstruct the signal exactly as it was sent.

digital audio tape (DAT) A method of recording information in digital form on a 5.25" audio tape cassette, developed by Hewlett-Packard and Sony. Approximately 1.3 giga-bytes of information can be recorded on a cassette using this method, compared with 300 megabytes on a normal cartridge backup tape.

digital circuit A transmission line that carries information in digital form, in other words as a series of coded electrical pulses. Digital information, such as a coded text message, can be carried on a digital circuit as it is, without the conversion by means of modems that is necessary to carry digital information over analog channels. Telephone conversations can also be carried on digital circuits. To do so, the analog voice signal is converted to digital form by sampling it at intervals, and converting the frequency found into a digital code – see *pulse code modulation*.

digital computer A machine that can accept a series of instructions (a program), and can then interpret these instructions while accepting and processing data, in order to generate new information or to control another machine or process. Both the program and the data are represented in digital form, as codes or numbers, and this distinguishes digital from analog computers, which operate on physical values such as voltages or movements. Most digital computers use binary notation, because this suits the semiconductor technology used to build their logic and memory circuits. Digital computers are extremely versatile and are used in a wide range of commercial, industrial and military applications, ranging from controlling fighter aircraft to processing the weekly payroll. They range in size from complete computers on an integrated circuit a few mm square, through personal computers that fit on a desk, to large-scale machines capable of storing and processing huge quantities of information.

digital imaging The creation of images of objects, using digital means to represent them. Since internal representation is digital, images can be processed, for example to rotate or tilt them or to view them from another aspect. Digital imaging is used in the medical field, such as by body scanners (see *computerised tomography*), in engineering and in mapping applications.

digital optical recording See *optical storage*.

digital paper A recently developed storage medium for digital information. It is a thin, shiny, flexible substance consisting of a polyester substrate with a metallic reflector and dye-polymer coating. It can be sliced into strips for tape drives or cut into disks for disk drives. Like conventional WORM (write once, read many times) optical disk drives, it uses optical techniques to record information and, once recorded, information cannot be erased and replaced. Like WORM again, digital paper can store very large volumes of information – 10 megabytes of data can be held on one square cm or 1 gigabyte on a 5.25" disk. But information can be accessed and transferred faster than from conventional WORM drives and storage costs are claimed to be lower – as cheap as paper, hence its name.

digital switching The establishment of connections within communications networks by processing the signals in digital form, rather than by altering the physical paths that the signals follow. The latter was the normal technology used for telephone exchanges until the 1970s, and is now commonly referred to as space switching. The most widely used technology for digital switching is known as time division multiplexing.

digital-to-analog conversion (D/A) The conversion of a series of numeric values into a continuously varying signal. Within a computer display, for example, digital-to-analog conversion is used to turn the digital representation of the contents of the screen, held within computer memory, into a varying signal that the electron gun(s) continuously trace across the screen to produce the image.

digital vascular imaging (DVI) A technique used to investigate blood circulation disorders. A liquid is injected into blood vessels to provide contrast when X-rays pass through it. The X-ray signals are analysed by computer, which builds an image of the blood vessel on a screen, showing details such as blood flow and blockages.

digital video interactive (DVI) A standard for recording video information on compact disk, originated by RCA and now owned by the chip manufacturer, Intel. It enables a developer of an interactive multimedia application to store about an hour of full motion video and audio on a CD-ROM, so that it can be read back and displayed under computer control.
    See also *compact disk interactive* (CD-I), a rival standard originated by Philips.

digitiser A device that converts analog information into digital form, so that it can be processed by a digital computer.

dimensioning See *sizing*.

DIP Abbreviation of *document image processing*.

DIP Abbreviation of *dual in-line package*.

direct access storage Storage from which data can be retrieved rapidly (in milliseconds), independently of the position that it occupies on the storage medium. Disk drives are the dominant example. Contrast with *serial storage* devices, such as tape drives, from which data can only be retrieved in the same sequence that it was previously written.

direct broadcasting by satellite (DBS) The broadcasting of television signals via satellite. Signals transmitted by the television station are bounced off a satellite, then picked up by roof-mounted dish aerials, from where they are carried to the set via wires or coaxial cable. The advantage of satellite broadcasting is that signals from a single television station can be broadcast to a very wide area, such as an entire continent. The area so covered is known as the footprint.

direct call An arrangement on a digital data network that enables a calling device to establish a connection very rapidly. When the network receives a call request, it interprets this as an instruction to make a connection with a device at a predefined address, without going through the normal preliminaries of call acceptance.

direct file See *random file*.

direct memory access (DMA) A method of transferring data from a peripheral to the memory of a computer system without the active intervention of the processor. Devices that operate in this manner 'steal' cycles from the processor to move data directly into or out of memory.

directive A statement in a programming language that tells the compiler what to do rather than forming a part of the program itself. For example, an EQUATE directive tells the compiler to treat one element within the program as equivalent to another.

directory A means of locating resources managed by a computer system or systems. The directory may be accessed explicitly by users, as is the case with a directory of the users of an electronic messaging service which holds both users' full names and the address used to send messages to them. Or it may be hidden from users and be used by the operating software to translate their logical requests into physical hardware addresses – for example, personal computer operating systems record the physical whereabouts of data files on floppy disks in a directory stored on each disk, but users and applications programs deal only with the names of the files.

directory service A service provided for the users of a computer network that enables them to address other users and services available via the network by meaningful names rather than by physical addresses. Also known as *name service*.
See also *X.500*.

disable Temporarily override or suppress some hardware or software feature. For example, a diagnostic routine might disable memory protection logic in a computer system so that it could test the operation of the memory.

Compare *enable*.

disassembler A program that translates a machine language (object) program into the assembly language equivalent.

disaster planning Preparations for surviving a disastrous event such as a fire or flood that puts completely out of action computer systems or supporting equipment such as commmunication networks. Disaster planning is becoming increasingly important as organisations come to rely more and more heavily on their computer systems for their continued operation, if not their survival. Measures that are adopted include establishing stand-by arrangements with other organisations, or installing duplicate systems at separate locations.

disc See *disk*.

disk A direct access storage device consisting of one or more flat circular disks, coated on one or both surfaces with magnetisable material. There is at least one read/write head for each surface, held on an arm which projects over the surface. Under the control of a computer, the arm can be positioned and used to record data magnetically, and to retrieve recorded data subsequently when required. Each surface is normally divided into tracks (a series of concentric rings) and each track in turn is divided into sectors (an arc of each track). Data is normally recorded as a series of 8-bit characters (bytes) each with a parity bit. Modern devices also reserve some spare tracks on the disk and bring these into use automatically whenever faults arise in other tracks on the surface.

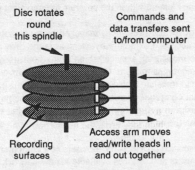

Disc rotates round this spindle

Commands and data transfers sent to/from computer

Recording surfaces

Access arm moves read/write heads in and out together

Disks fall into two main categories – hard disks, whose recording surface is rigid and where the capacity of one disk ranges from about ten up to thousands of megabytes; and floppy disks, with a flexible recording surface, found on word processors and personal computers, with capacities from one hundred or so kilobytes up to about one and a half megabytes. They may also be either exchangeable, in other words removable from the drive, as is the case with floppy disks and many hard disks, or permanently fixed within the drive casing, as is the case with the Winchester disks used on personal computers.

disk cartridge A disk storage medium in cartridge form, so that it can be removed and stored offline after processing.

disk controller A device that controls one or more disk drives. It receives commands from the computer to which it is attached, forwards these to the drives under its control, supervises their execution of the commands and returns data and status information to the computer. Usually, it will also contain routines to help with recovery from hardware failures, for example bringing spare tracks into use automatically when other tracks become unusable, and diagnostic routines for use by maintenance engineers.

disk drive A computer peripheral that is used to store and retrieve information held on disk media. Small disk drives such as floppy or Winchester disks include both the read/write mechanisms, used to record and read information on the disk, and the logic that controls the drive and enables it to communicate with the computer system to which it is attached. Larger disk drives such as are used on mainframe computers are connected to the computer via a disk controller that contains most of the control logic. A number of drives can be connected via each controller.

disk mirroring The technique of writing data to two separate disk storage subsystems simultaneously. This is a common way of making a computer system fault tolerant, since it can continue without interruption if one of the disk subsystems fails. Disk drives are one of the least reliable components of modern computer systems.

disk operating system (DOS) An operating system designed to take advantage of disk storage devices. Such an operating system loads itself into memory from disk automatically when the computer system is started up; may also load additional routines from disk when they are needed; and usually includes routines which enable applications programs to store and retrieve data held on disk.

disk sector An arc of a track on a disk, which can be read or written in a single operation. Before they can be used to store information, disks normally have to be initialised by the computer system on which they are to be used. During the initialisation process, markers are written on to the disk to show where sectors start.

disk server See *server*.

diskette See *floppy disk*.

display PostScript See *PostScript*.

87

display screen The component of a visual display terminal that displays information. Except where space is at a premium, a cathode ray tube (hence the alternative term – CRT) is used, using similar technology to a domestic television. For small screens or where low power consumption is important, as for portable terminals, *liquid crystal* or flat-panel *plasma* displays are used instead, while very small screens that must be visible in darkness sometimes use *light emitting diodes*.

display size The size of a display screen measured diagonally from corner to corner. In the UK and the US this is normally expressed in inches, personal computer screens varying from a minimum of 9 or 10 inches up to 26 inches for desktop publishing.
   See also *resolution*.

display text Text in large sizes, usually above 14-point.
   Compare *body text*.

distributed computing See *distributed data processing*.

distributed data interface (DDI) A standard prescribing how a data network should work and what it should consist of, ratified by ANSI in late 1990. ANSI's main purpose was to ensure high performance and good fault tolerance, and the basic arrangement consists of two rings round which data flows in opposite directions at 100 megabit/s, controlled by a token passing protocol. Individual machines on a network can be up to 2 km apart and the whole ring can be up to 100 km in length. It can be implemented using either optical fibre (see *FDDI*) or copper cable. The standard defines components, speeds and configurations conforming to the two lowest layers in the open systems interconnection (OSI) model – the physical and data link layers.

distributed data processing (DDP) Data processing where a number of independent computer systems are installed at different locations to process and store data that originates locally. These systems also form part of an overall system, and exchange data with one another when necessary, either via a communications network or by exchanging disks or tapes. The advantage of distributed processing is that the departments or branches in which the computer systems are installed can control their own processing, but the overall arrangement is more difficult to manage than a centralised transaction processing system, because of the need to co-ordinate the updating of data files at the various locations.

distributed database A database whose component files are spread across computer systems at more than one location. Assuming that the database is a true database, or in other words that the relationships between records in the database are recorded within the files, then the database management system handling them has to keep relationships up to date in all the component files when records are changed. Sometimes this means making changes to database files stored at locations other than where a change originates. To do this, and also to assemble answers to queries which draw data from more than one location, the software has to maintain a directory specifying where data items are stored. If files are badly designed, distributed databases can run into serious problems with performance and with database recovery.

distributed processing See *distributed data processing.*

dithering A technique used by some colour painting applications on personal computers. It involves mixing pixels of different colours and densities into clusters that the eye perceives as new colours.

DMA Abbreviation of *direct memory access.*

DML Abbreviation of *data manipulation language.*

DO statement Part of a control structure used in some high-level languages to construct loops within a program.
See also *FOR statement; WHILE statement.*

docking station A cradle into which a portable computer is placed so that it can be used as a normal desktop computer, i.e. with a high-resolution screen, with network links to other computers, etc.

document content architecture (DCA) A format for the content of documents to be exchanged between dissimilar computer systems, defined by IBM. It includes in particular a revisable text format. As well as the normal printable characters, this specifies standard codes for indentation, tabs and other text formatting information. This means that the document can be displayed and edited on the receiving system in exactly the same form that it assumed on the sending system.
See also *document interchange architecture.*

document image processing (DIP) The capture, storage and processing of documents in the form of digitised images, rather than as coded data or text. Documents are scanned to create the image, which is normally stored on a large-capacity optical storage device. After they have been scanned and stored, identifying information is associated with each of the images, so that they can be retrieved later for examination or further processing. Document images require substantially more storage space than if they are keyed in as text – 100,000s of bits per A4 page as against only a few thousand. The advantages of storing them in this form are, first, that any form of information can be captured, rather than just what can be typed on a keyboard – drawings, photographs, signatures, etc. Secondly, less labour is needed to capture them, since the scanner does most of the work of encoding.
Document image processing is an alternative to filing on paper or on microfilm. It is

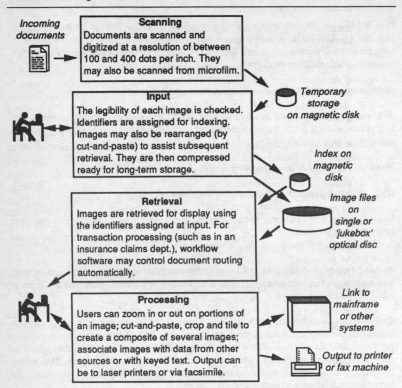

Incoming documents

Scanning
Documents are scanned and digitized at a resolution of between 100 and 400 dots per inch. They may also be scanned from microfilm.

Temporary storage on magnetic disk

Input
The legibility of each image is checked. Identifiers are assigned for indexing. Images may also be rearranged (by cut-and-paste) to assist subsequent retrieval. They are then compressed ready for long-term storage.

Index on magnetic disk

Retrieval
Images are retrieved for display using the identifiers assigned at input. For transaction processing (such as in an insurance claims dept.), workflow software may control document routing automatically.

Image files on single or 'jukebox' optical disc

Processing
Users can zoom in or out on portions of an image; cut-and-paste, crop and tile to create a composite of several images; associate images with data from other sources or with keyed text. Output can be to laser printers or via facsimile.

Link to mainframe or other systems

Output to printer or fax machine

cheaper than paper, reducing both staffing and space requirements, and reducing the risk of misfiling or loss. Compared with both, it enables documents to be retrieved much more rapidly and also permits multiple access over geographical distances, thus eliminating the need to duplicate documents or transport them physically from place to place.

document interchange architecture (DIA) A set of rules for the electronic exchange of documents prepared and stored on computer systems, drawn up by IBM.

See also *document content architecture*.

document management The methods used to organise the storage and retrieval of documents, in other words information consisting mainly of text, held in computer files. Document management differs from data management (and database management) in that much less precise criteria are used to identify and search for information. A typical data query might be 'Tell me the total sales of widgets in Brazil in March 1988', whereas a typical document query would be 'Find me all the documents which say anything about the Brazilian economy in 1988'.

document routing A feature of some document image processing systems, aimed at controlling the route that a document follows through its various stages of processing. In an insurance office, for example, a system might route incoming mail to the appropriate clerk for data entry; refer queries or sensitive transactions to the supervisor; suspend processing of a document if further information was needed; and so on.

document scanner A device used to scan documents and to represent their contents as digital signals that can be transmitted or processed by computer. To scan a document, it is divided into a series of discrete elements or pixels (short for picture elements). Arrays of either photodiodes or charge-coupled devices (CCDs) measure the lightness or darkness of each pixel of the document and convert this into a numeric value, doing this line by line, at a resolution of between 200 and 400 pixels per inch. These numeric values are then read and used to build a digital representation of the document image. Information in this form requires much more storage space than a coded representation of the characters, such as produced by a word processing program. After it has been compressed, a scanned A4 page requires about 50 kilobytes of storage, as compared with about 2.5 kilobytes for a word processed page. Expensive high-speed scanners can process about one page per second, whereas cheaper desktop scanners manage about ten pages per minute.

Similar technology is used for *optical character recognition*, with the addition that character images are converted into their equivalent character codes by software after the initial scan.

documentation The documents that describe a computer application and the individual programs it comprises. Documentation is normally produced by those who design the application and write the programs. Its purpose is both to enable intended users to operate the programs successfully, and also to enable maintenance programmmers or others (who will probably not be the originators) to change the way they operate in the future. For a large application such as might be developed by a computer department, the documentation normally includes:

(1) a system specification, usually describing the system both in functional (what it does) and technical (how it does it) terms;

(2) program specifications for each of the programs;

(3) data or file specifications for any shared data files;

(4) user documentation such as operating instructions, details of backup procedures, and so on.

For a program developed on a personal computer, the documentation might consist of a description of the function of the program; operating instructions, including the formats of screens used for input and of printed reports; data file formats; and a program listing.

domain (1) A column in a relational database file, in other words all the values of one particular attribute. This is analogous to a field in a conventional data file.

(2) The boundary set for the knowledge base of an expert system, usually described as the problem domain.

dongle A small hardware device that can be plugged into a computer, and without which it will not run a particular piece of software, such as an applications package. The dongle is supplied to those who purchase the software, and is a way of preventing illegal copying of the software – see *software piracy*.

dopant The impurities introduced into semiconductors in order to give them the electrical characteristics required for integrated circuits. Typically, the materials used are phosphorus and boron. They alter the electrical properties of the silicon in a special way. Pure crystalline silicon is a good electrical conductor, but when 'doped' with these impurities shows either an excess (n-type, for negative) or a deficiency (p-type, for positive) of electrons when a voltage is applied. If a sandwich (n-p-n or p-n-p) of these materials is created, the current flow between two of the elements can be regulated by a voltage applied to the third. This is the principle of the transistor.

DOS Abbreviation of *disk operating system*.

dot matrix A method of forming characters out of a grid of dots, used by printers and on visual displays. Each character is formed by using a particular combination of dots. Inexpensive printers normally use a matrix 9 dots by 9, but more expensive so-called 'near letter quality' matrix printers may use an 18 by 18 dot matrix.

dot matrix printer See *matrix printer*.

dots per inch (dpi) A commonly used measure of the resolution of a display screen or a document scanner. In both cases, the image is formed by means of a grid of dots. A typical scanner resolution is 200 dots (or pixels) per inch, while a typical graphic display has a resolution of about 70 dots per inch.

double buffering A method of handling blocks of data read from or written to mass storage files that are processed serially, designed to accelerate processing. Two buffers are allocated in computer memory for each file. While an applications program is taking input records from or adding output records to one of the buffers, data is being transferred between memory and the mass storage device to or from the other buffer. This makes it possible to overlap processing of records with transfer of data to and from the mass storage device.

double-precision arithmetic Arithmetic carried out using double words (i.e. two adjacent words) for the operands, so that larger numbers can be handled and/or results are more accurate.

double-sided Used of floppy disks, both sides of which can be used for data storage.

DOV Abbreviation of *data over voice*.

down time The time that a computer system is not available for service. This may be as a result of system failure or recovery from failure, and for scheduled and unscheduled maintenance and repair work.
Compare *idle time*; *up time*.

download Transfer a program or a file from one system or device in a network to be processed by another lower in the control hierarchy. Thus a data file might be downloaded from a mainframe computer to a personal computer, and a document file might be downloaded from a personal computer to a laser printer.

downsize Substitute smaller (and, in aggregate, cheaper) computers for a larger one, such as several minicomputers for a mainframe computer.

downwards compatible Useable with less powerful or earlier versions of a product. The term is most often used to describe new versions of software such as operating systems or applications packages. If these are described as downwards compatible with the previous version, it means that procedures and programs developed to operate with that earlier version will continue to operate properly, without modification, with the new version.

DP Abbreviation of *data processing*.

dpi Abbreviation of *dots per inch*.

drag Used to describe an operation performed with a mouse, in which the cursor is positioned using the mouse, the mouse button is pressed, then the cursor is moved to another position with the mouse button held down. Dragging is used to move objects about and to draw shapes on the screen.

DRAM Abbreviation of *dynamic RAM*.

driver A program that handles the basic operations of a peripheral or a transmission line, such as initiating physical transfers, checking transfer status, and initiating repeat transmissions on detecting hardware error. Drivers may be held in read-only memory (ROM) or on expansion cards, or may be included within the operating system.

drop A point on a multipoint line at which the line branches in two directions, either to connect a terminal or to continue the line to the next drop.

drum storage See *magnetic drum.*

dry running See *desk checking.*

DSS Abbreviation of *decision support system.*

DTE Abbreviation of *data terminal equipment.*

DTMF Abbreviation of *dual tone multi-frequency.*

DTP Abbreviation of *desktop publishing.*

dual in-line package (DIP) An integrated circuit packaged so that it can be plugged into a printed circuit board through two parallel lines of pins. DIP circuits are commonly used for switches that can be altered by the user to set the operating characteristics of peripherals. Character printers, for example, often have DIP switches to set for US size or A4 paper; for different national character sets; for varying transfer speeds to the computer, and so on.

dual tone multi-frequency (DTMF) The technology used in pushbutton telephones, commonly known as touchtone, of generating audible notes to represent numbers.

dumb terminal A terminal which cannot store or manipulate the data it receives, but can only send it directly to a display or printer. Contrast with *intelligent* (or *smart*) *terminal*, which can store data and process it before it is displayed or printed, responding to commands included in the data or using a program stored in the terminal itself.

dump Transfer to another medium, usually as a complete record. See *memory dump*, *file dump*, *incremental dump* or *snapshot dump*. Sometimes also used as synonymous with *checkpoint*.

duplex See *full-duplex.*

DVI Abbreviation of *digital vascular imaging.*

DVI Abbreviation of *digital video interactive.*

DXF A proprietary format for files representing CAD images, defined by Autodesk and incorporated into the company's widely-used CAD package, AutoCAD. It is effectively a standard for the exchange of two-dimensional CAD images, widely supported by CAD packages for importing and exporting information.

dye sublimation A technology for colour printing capable of producing continuous-tone images. Plastic film coated with dye in the three primary colours plus black (see *CMYK*) is passed across a print head containing several thousand heating elements. The elements can produce 255 different temperatures and the hotter they are the more dye is transferred. Coated paper, designed to absorb the gaseous dye on contact, passes across the film four times, once for each colour, as it passes the print head.

Compare *phase change*.

dynamic allocation Allocation of computer system resources, such as peripherals or space in memory, as and when they are needed by a program, rather than by reserving them explicitly when a program is loaded. Also known as *adaptive allocation*.

(1) Of peripherals – programs running in a large multiprogramming system specify the identity of tapes and disks they wish to process, but not the drives on which they are to be mounted. When a program calls for a particular tape or disk file (a logical unit of storage) or volume (a physical unit of storage), the operating system searches for it on all the available drives. Having found it, the file or drive is allocated to the program concerned, and all subsequent reads and writes addressing the file/volume are directed automatically to the correct drive.

(2) Of memory – the operating system allocates areas of memory to programs when they are loaded and when they request additional space to hold data. If there is competition from programs for the available memory, the operating system will allocate it to programs according to their priority and other factors, so as to use it as effectively as possible.

See also *memory management*.

dynamic binding Where the connection between the name of an operation and the program code to carry out the operation is established at run time, rather than when the program is compiled. This means that libraries of shared routines can be changed without affecting the programs that use them.

Compare *static binding*.

dynamic memory See *dynamic RAM*.

dynamic RAM (DRAM) The cheapest form of random access memory made from semiconductors. It is called 'dynamic' because it loses its contents by leakage and every time it is read, so that it has to be refreshed about every two milliseconds and also, of course, after it has been read.

Compare *static RAM*.

dynamic range The ability of an input or output device (for example a scanner) to distinguish or reproduce subtle differences in grey shades or colours.

DYUV Abbreviation of *delta YUV*.

E

EAROM Abbreviation of *electrically alterable read-only memory*, sometimes called the 'read-mostly' memory. It is used for applications requiring a memory that can be altered, but where read operations are much more frequent than write operations. Unlike the EPROM, where the entire contents of the memory is erased before a new pattern of information may be imposed, the EAROM can be altered selectively. It uses special materials during its fabrication to build small circuits that may be selectively charged and discharged. Each such cell will hold a charge until it is erased by a strong pulse of current, after which a new charge may be electrically imposed on the cell while leaving other cells unchanged.

EBCDIC Abbreviation of *Extended Binary Coded Decimal Interchange Code*. A character set (i.e. the codes identifying different characters) defined by IBM for its Series/360 and subsequent ranges of general-purpose computers. It uses 8 bits for each character, giving 256 unique combinations. It is used both for internal processing and to transmit information between computers and terminals.

echo A news area on a bulletin board system, identified by topic. Subscribers dial in to the bulletin board, specify which echo they are interested in, and can then read the latest news items – messages left by other subscribers – in chronological sequence. If they wish, they can enter a message themselves, containing news of interest to subscribers or comments on other messages.

echo check See *echoplex*.

echoplex A procedure used with asynchronous terminals to check that data has been transmitted correctly over a communications link. When sending a message, the transmitting terminal does not print out or display the characters that are being transmitted. The receiving terminal (or sometimes the network node) sends back (echoes) the characters it receives, and the sending terminal prints or displays these when they arrive. Thus the user at the sending terminal sees on the printed log or on the display what has been received rather than what has been sent, and can verify visually that this is correct.

ECL Abbreviation of *emitter coupled logic.*

ECMA Abbreviation of *European Computer Manufacturers' Association.*

edge connection socket See *expansion slot.*

EDI Abbreviation of *electronic data interchange.*

Edifact Abbreviation of *Electronic Data Interchange For Administration, Commerce and Transport,* a set of syntax rules for messages used to exchange trading data between business organisations. Edifact is a draft international standard jointly defined by the US and Europe.

editing terminal The terminal used by an information provider to input data – frames and routing information – into a videotex system.

editor A program provided as part of an operating system or with a programming language that allows the user to enter and amend source programs or similar text files. More limited than word processors in terms of formatting features, normally producing files in text only format.

EDP Abbreviation of *electronic data processing.* Synonymous with data processing.

effective data transfer rate The speed at which data bits are transferred along a transmission channel, excluding any additional information used to detect and correct errors. Often extra bits are added as a redundancy check, and blocks will also be retransmitted occasionally when transmission errors are detected. The effective data transfer rate is normally expressed as the average number of bits (or kilobits or megabits) transferred per second.

EFT Abbreviation of *electronic funds transfer.*

EFTPOS Abbreviation of *electronic funds transfer at point of sale.* See *electronic funds transfer.*

EGA Abbreviation of *enhanced graphics adaptor.*

egoless programming Where programmers suppress their natural pride in their own creativity in the common interest. Coined to describe an aspect of structured programming which requires staff to check one another's work in order to avoid idiosyncratic programming techniques that might lead to difficulties with maintenance.

EIA Abbreviation of *Electrical Industries Association*.

8-bit (1) Used of a microprocessor or a bus (the data 'highway' within a small computer) to specify how many bits it can handle at once. Together with speed of operation (expressed in MHz) this is an important indicator of performance. (2) Also used to specify how many bits are used to achieve a particular effect, as in '8-bit colour'. Since 8 bits can carry 2^8, i.e. 256, combinations, 8-bit colour display screens are capable of displaying 256 different colours.

80-column display A visual display where the screen is divided vertically (80 columns) and horizontally (usually 24 or 25 lines) into invisible cells, in each of which an alphanumeric character can be displayed, using a character generator built into the display. 80-column displays can be contrasted with *bit-mapped displays*, which are divided into a grid of tiny dots. Bit-mapped displays have the advantage of great flexibility, being able to display a combination of text and graphics, but place greater demands on the processor and the software driving them.

EIS Abbreviation of *executive information system*.

EISA Abbreviation of *extended industry-standard architecture*.

Electrical Industries Association (EIA) A US trade association that has originated a number of interface standards for telecommunications. See *RS series recommendations*.

electro optics See *fibre optics*.

electronic blackboard An image drawn by hand on a writing tablet at one site is condensed, transmitted and reconstituted at another site. The user at the second site can erase and add to the image seen on his monitor and send a condensed and reconstituted version back to the original user's monitor. This process is so fast that the interactive process is perceived as though the participants were seated side by side.

electronic cottage The concept of the home as a workplace, linked electronically to colleagues and customers based elsewhere.

electronic data interchange (EDI) Electronic data interchange has been defined (by the International Data Exchange Association, based in Brussels) as 'the transfer of structured data, by agreed message standards, from computer to computer, by electronic means', but the term is most often used in a narrower sense, to mean the exchange of trade transactions such as orders and invoices between trading partners, usually between supplier and customer or distributor. The aim of EDI is paperless trading, and an end to postal delays

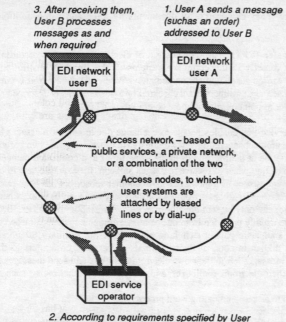

3. After receiving them, User B processes messages as and when required

EDI network user B

1. User A sends a message (suchas an order) addressed to User B

EDI network user A

Access network – based on public services, a private network, or a combination of the two

Access nodes, to which user systems are attached by leased lines or by dial-up

EDI service operator

2. According to requirements specified by User B, Service Operator either stores the message in User B's mailbox to await collection or forwards the message immediately

and the duplication of effort involved in re-keying documents generated by computer in the first place. EDI is used within a growing range of industries, including financial services, transport, shipping, food, retailing and manufacturing.

electronic data processing See *data processing*.

electronic document distribution The controlled distribution of documents by electronic means. Typically, images of documents are stored centrally and lists of documents relevant to particular individuals are sent out periodically. Individuals can then inspect the document images on display screens and/or request that a version be sent to them on paper.

electronic filing The use of a computer system for office filing. Normally, an electronic file is shared by a number of people in a work group. Each has private space in the file and can also add material – documents, notes, messages, perhaps 'voicegrams' – to shared files on various topics. They can search for documents by topic and using multiple keywords.

electronic funds transfer (EFT) The transmission of electronic messages recording financial transactions directly to the computer systems of the financial institutions involved, so that they can be accounted for immediately. Where the transaction is captured at point of sale, such as at a supermarket checkout or in a store, this is described as 'electronic funds transfer at point of sale' or, more commonly, EFTPOS.

electronic mail A service that enables people to exchange documents or messages in electronic form. These are exchanged non-synchronously (in other words the two parties to an exchange need not be at their terminals at the same time) in a controlled manner. Users of a service can decide when to receive messages and can, for example, ask to be sent an acknowledgement as soon as a message sent to another subscriber has been read. It is this which sets electronic mail apart from established services such as telex, where messages (by and large) can only be sent directly from one telex station to another. Electronic mail services are usually provided from one or more computer systems attached to a switched network, such as a packet-switching network or the ordinary telephone network. Subscribers dial in to connect their terminals or personal computers to the service, then can read messages left in their own electronic mailbox and send messages to other subscribers via their electronic mailboxes, and sometimes by other means such as telex or fax.

Formerly referred to as communicating word processors.

electronic mailbox An area of storage within an electronic mail system, designated to hold mail addressed to a particular subscriber. Messages addressed to the subscriber are put into this electronic mailbox. When the subscriber connects to the system, s/he can look through the messages in the mailbox one by one and decide what to do with them. The system usually provides options such as 'reply' (i.e. put a reply message into the originator's mailbox), 'forward' (i.e. send on this message to another subscriber along with a covering message), 'delete' or 'leave'.

electronic messaging A general term covering all forms of electronically-mediated communication. This includes electronic mail for text messages and an equivalent service that uses recordings of spoken messages, known as voice messaging. It may also include computer conferencing and videotex. Also used as synonymous with electronic mail.

electronic office One of many terms coined in the late 1970s to describe the office following the introduction of information technology. Other favourites were 'office of the future' and 'paperless office'. In practice, the office still remains dominated by people and paper rather than electronic equipment, although the number of the people needed to get through the work has begun to decline and the quantity of the paper has if anything increased.

100

electronic point of sale (EPOS) Describes equipment used to capture details of transactions at the point where a sale takes place, such as at a supermarket checkout or in a shop. This includes electronic cash registers and scanners used to read bar code labels.

electronic publishing See *desktop publishing*.

electronic shopping See *teleshopping*.

electronic wand A small hand-held device that is used to read printed marks such as bar codes on products or labels. The head of the wand is passed across the material to be read and sends the appropriate information to the computer to which it is attached.

electrostatic printer A non-impact printer in which electromagnetic pulses and heat are used to imprint characters on the paper.

element Any item of data that can be treated as a unit in a particular situation, such as the individual fields in an array or a record.

elementary item Used in COBOL programming to mean a data item with no subsidiary items, in other words a data field described at the lowest level of detail.

ELSE statement See *IF statement*.

email See *electronic mail*.

embedded Integrated into another device or routine in such a way that the user of the latter is not aware of its presence. Used, for example, of computers that serve as control components in electromechanical devices like washing machines or cameras.

emitter coupled logic (ECL) A high-performance integrated circuit technology, used in the processor logic of powerful mainframe computer systems.

empty shell See *expert system shell*.

emulate Operate in the same manner as, from the point of view of other system components with which the emulating device interacts. Used of devices that can appear to other devices to behave as a different type of device, usually by running special emulation software. Personal computers, for example, can emulate visual display terminals in order to retrieve data from mainframe computers, and some data terminals can emulate a number of different types of terminal so that they can work with the widest possible range of computers.
  Compare *simulate*.

enable Restore or activate some hardware or software feature that was previously disabled.

encapsulated PostScript format (EPSF) A format for files recording graphic images that are to be printed on devices (such as laser printers or phototypesetters) using the PostScript page description language. It includes the PostScript program instructions that the printing device will interpret to recreate the graphic, plus additional 'encapsulation' giving the dimensions of the image and, optionally, the image in bit map form (for which representation standards may vary). This means that graphics software such as desktop publishing programs can work with the image (or a rough version of it) without needing to interpret the PostScript program. EPSF is relatively inefficient as a means of representing bit map images, for which TIFF format is also widely used.
Compare *tagged image file format*.

encapsulation Packaging up a set of properties and behaviour so that an external observer (the potential user) can easily deduce what are the intended purpose and benefits of the package, without needing to be aware of any of the internal detail. The encapsulated package may include whatever mix of hardware and software, program and data, etc. are necessary.
See also *information hiding*; *object orientation*.

encryption The encoding or scrambling of data prior to transmission so that it cannot be recognised if intercepted during transmission. The reverse process (decryption) is carried out on receipt so that the data can be processed.
See also *data encryption standard*.

encyclopaedia An extended form of data dictionary. As well as details of the attributes of data items used by applications programs normally held in a data dictionary, it also defines other aspects of the way applications programs operate, such as screen formats, the contents of menus and other standardised procedures.
See also *repository*.

end of file block See *end of file label*.

end of file label A qualifier block on a magnetic tape which marks the end of a file.

end of file routine A routine within a program that is to be executed when the end of a data file is reached. Usually, the name of this routine is specified when the file is opened. If no name is provided, a standard routine included in the operating software is used.

end of page routine A routine within a program that is executed whenever there is no more space on the current page of a document that is being printed. Its job is to print any trailer information (such as sub-totals or page number) on the current page, then move the paper to the head of the next page and print any headings.

end of reel block See *end of reel label*.

end of reel label A qualifier block on a magnetic tape which marks the end of a reel, implying that the file continues on a subsequent reel.

end of tape marker A reflective strip on a magnetic tape, marking the physical end of the recording surface on the tape.

end of text (ETX) A control character used to mark the end of the text of a message. If the message is sent in a series of transmission blocks, the last will terminate with an ETX character, and all the others with an ETB (end of transmission block) character.

end of transmission (EOT) A control character used to mark the end of a transmission of data. For example, an EOT character might be sent after all the data in a file transfer has been sent. On receiving it, the called terminal knows that it can close the file and, if it wishes, clear down the call.

end of transmission block (ETB) A control character used to mark the end of a transmission block. If a message is sent in a series of transmission blocks, the last will terminate with an ETX character (end of text), and all the others with an ETB character.

END statement A statement used in some high-level languages to mark the end of a program or the end of a block (see *block-structured lanaguage*) or the end of a compound statement.

end-to-end control Control of the accuracy and the flow of data all the way between originating and receiving devices. Computers normally apply end-to-end control to the flow of traffic between themselves and their terminals, initiating retransmission when necessary. The components which make up the communications link over which the traffic is flowing will control the reliability of transmision over each portion of the link separately.

end-user The ultimate user of a computer system, rather than the people that purchase it or develop applications programs for others to use. In the computer industry, the term 'user' originally meant the purchaser, which was usually the data processing department of an organisation. Another term was needed for those people in the organisation who were the real users, in the sense that the computer system ran applications programs, generated reports, and so on, to help them do their jobs. 'End-user' has since been adopted with precisely that meaning. In terms such as 'end-user computing', it emphasises the point that end-users no longer rely completely on the specialists for a kind of chauffeur service in order to get useful work out of a computer system. Given the right tools, now widely available, they can sometimes drive it themselves.

end-user computing Computing where the end-user decides how and when processing should take place. End-user computing really began with timesharing, which allows end-users, working at terminals, to build and run applications on a large shared computer system, but the term is most often applied to applications on personal computers which are set up (by programming or by some other means) and run by end-users themselves. Examples are spreadsheets, or simple data processing applications built using application generators such as dBase.

end-user systems Computer systems where the end-user exercises much of the discretion as to when and how processing should take place, rather than the computer specialists or the systems themselves. This includes end-user and departmental computing, and office systems for such applications as electronic filing and messaging. End-user systems can be contrasted with *operational systems*, which handle the basic processes of an organisation.

engineering workstation See *workstation*.

enhanced entity-relationship model See *semantic data model*.

enhanced graphics adaptor (EGA) A standard for the display of information on the screen of a personal computer, and also a plug-in expansion card that implements the standard. It provides for both text and graphics, implemented in separate modes. In text mode, characters are constructed from an 8 x 14 matrix of dots. There are three graphics modes: medium (four colours, with a resolution of 320 x 200 pixels); high (16 colours, 640 x 200); and ultra high (16 colours, 640 x 350 pixels). The last is intended for applications such as computer aided design and desktop publishing.

enhanced small device interface (ESDI) An interface used for connecting peripherals such as disk drives to small computers.

ENQ Abbreviation of *enquiry character*.

enquiry character (ENQ) A control character used to find out whether a terminal is ready to send or receive data. For example, ENQ characters are sent out by a computer system or a network processor using poll/select procedures over multipoint lines. On receiving an ENQ addressed to it, a terminal knows that the line is free for it to send a message to the computer system.

enterprise model See *corporate data model*.

enterprise schema See *conceptual schema*.

entity Anything about which you need to gather information in order to construct an information system. Entity is a term particularly used in a database context to describe the tangible objects (such as products or customers), events (such as purchases) and abstract concepts (such as product groups) which are modelled in a database and in the applications that process the information stored there.

See also *entity-relationship model*.

entity life history A systems analysis technique based on identifying the different states that an entity may go through during its life, and the events and processing functions that cause these to change. A student, to take a simple example, begins as an applicant, then is registered as a student, and finally becomes a graduate. The entity life history is represented in a diagram (the figure shows some typical symbols used), and can be used to form the outline design of a computer application.

entity model See *entity-relationship model*.

entity-relationship model A form of data model, originated by Peter Chen, that describes data in terms of entities and the relationships between them. Entities are tangible objects (such as products or customers), events (such as purchases) and abstract concepts (such as product groups) with which a system deals, defined as having certain attributes or properties (names, codes, numeric values, etc.) and certain relationships with other entities. As the very simple example below shows, doctors treat patients and run clinics, which patients attend. An entity model is easily understood, both by computer specialists and by end-users, and for this reason is often used to build a conceptual data model. This is used both for initial planning of information systems and, at a greater level of detail, as a means of designing database files. Sometimes referred to just as entity model.

See also *semantic data model*.

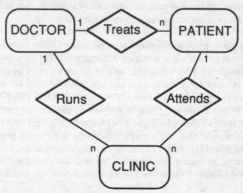

entity subtype By declaring an entity type as a subtype of another, attributes or relationships of the latter are automatically inherited by the former. Thus, for a personnel application both 'employee' and 'applicant' might be declared as subtypes of the entity 'person'.

entity/function matrix A diagram that shows in which functions of an organisation or computer application entities (types of record in a database) are handled.

envelope Those elements of a message sent over a transmission link that are there to facilitate transmission, as opposed to its information content (the text or data).

EOT Abbreviation of *end of transmission*.

EPOS Abbreviation of *electronic point of sale*.

EPROM Abbreviation of *erasable programmable read-only memory*. It is used for applications requiring a memory that can be altered, but where read operations are much more frequent than write operations. A store of charges on pairs of electrodes is built up to form the desired pattern of information, and will remain reliably in place for years. To alter the pattern, the entire contents of the memory is erased by exposing the chip to ultra-violet radiation, which allows the charges to leak away rapidly. After this erasure, a new pattern of information may be imposed.

Another type of EPROM is under development, known as the flash chip or flash EPROM, that uses a pulse of electricity to erase memory contents rather than ultraviolet light. It has price/performance characteristics approaching those of hard disks.

EPSF Abbreviation of *encapsulated PostScript format*.

EQUATE directive A directive used to indicate to a compiler that an identifier is to be treated as equivalent to a particular value or to another identifier.

erasable optical storage Optical storage on which data can be erased and rewritten by the computer system to which it is attached. Compare *WORM* (Write Once Read Many times) optical storage, on which data can only be recorded once by computer. WORM devices tend to be used for archival storage only, whereas erasable optical stores can substitute for disk storage, although access times are relatively slow. The most popular technology for erasable optical disks is known as thermo-magneto-optical or just magneto-optical. It uses a disk with a glass or plastic base coated with rare-earth materials. After formatting, the disk has a magnetic field in a direction perpendicular to the disk. This magnetic field is resistant to change at room temperatures, but when a tiny part of the disk is heated by laser a few hundred degrees, the field can be bent by electromagnet, indicating a binary 1, or returned to its original position, indicating 0.

ergonomics See *human factors engineering*.

106

erlang A measure of the traffic flow on a transmission line or through a network component. Erlangs, roughly speaking, represent the number of calls likely to be in progress at a particular time.

error Any instance where the expected results are not achieved. Errors with computer systems can be divided into three broad categories:

(1) hardware errors, resulting from a malfunction in the equipment;

(2) software or program errors, arising because those who specified, designed or wrote the program concerned made mistakes;

(3) operator or user errors, where an individual using a system has entered incorrect data, supplied the wrong instructions, or taken inappropriate action.

error code A code identifying the nature of an error that has occurred. It may be displayed or printed out so that a user can look it up in a manual, or may be used by an error routine in a program to decide what action to take.

error list See *error report*.

error rate See *bit error rate*.

error report A printed list of the errors detected during the execution of a program, or after a diagnostic test.

error routine A routine that is to be executed whenever an error condition or an unexpected event occurs. The name of an error routine may be specified in a source program at the head of a section of program or when a data file is opened, and the compiler arranges that it will be called automatically when appropriate. It simplifies the task of the programmer in providing for conditions that are unlikely to occur or only occur very rarely.

escape character A control character generated by pressing the Escape key on a computer or terminal keyboard, or sometimes a combination of keys, although it is most often generated automatically by software. It is used to signal to a receiving device that the character or characters following it have a special meaning. Together with the escape character, these characters are known as an escape sequence.

escape sequence A sequence of characters beginning with an escape character. Escape sequences are used to extend the range of a character set. For example, they are included in the stream of characters sent to a daisy wheel printer to specify the spacing between characters or lines, and to switch on or off special effects such as boldface or underline.

ESDI Abbreviation of *enhanced small device interface*.

ESPRIT programme A programme of collaborative research into information technology involving the member states of the European Community. Its aims are to promote transnational co-operation; provide European industry with the technology it needs to compete in the 1990s; and to contribute towards the development and implementation of international standards. ESPRIT phase 1 had a budget of 1.5 billion ecu and took place between 1983 and 1986. Research was divided into five areas: advanced electronics, software technology, advanced information processing, office systems and computer integrated manufacturing. ESPRIT phase 2 has a budget of 3.2 billion ecu and began in 1988.

ET1 A benchmark which is a cutdown version of Debit/Credit, allowing performance to be measured independently of the communications-related components, such as the teleprocessing monitor. Transactions are generated and fed to the computer system whose performance is being measured by an external driver system. TP1 is similar, except that the transaction generator runs on the computer system under test. Performance figures based on ET1 or TP1 are widely quoted by suppliers of database management systems.

ETB Abbreviation of *end of transmission block*.

etching A stage in the manufacture of integrated circuits where acid is used to remove unwanted layers of material.

Ethernet A type of local area network, originated by Xerox Corporation in 1975 and developed by Xerox, DEC and Intel. It has since been defined by ISO standard 8802/3 and adopted by a large number of other manufacturers. Ethernet networks operate at a speed of 10 megabit/s. They use either special low-impedance coaxial cable laid in a tree configuration or alternatively, in a subsequent development, twisted-pair wiring connected via a concentrator. Fibre optic versions are also under development.

Several hundred devices can be connected to a single length of cable. The capacity of the cable is shared among these devices by a form of protocol known as *carrier sense multiple access with collision detection* (CSMA/CD).

See also *10BaseT*; *thick Ethernet*; *thin Ethernet*.

ETHICS A method for designing computer systems, devised by Enid Mumford of the Manchester Business School. ETHICS stands for Effective Technical and Human Implementation of Computer-based Systems, and places particular emphasis on social and organisational factors affecting design, and on the participation of users in the design process.

ETX Abbreviation of *end of text*.

EUREKA programme A programme of collaborative research involving the member states of the European Community. Its aims are to remove technical barriers to trade and open up the public procurement system, prior to the creation of a single European market in 1992. As well as information technology, it covers robotics; biotechnology; materials, energy and transport technologies; manufacturing; and environmental protection.

European Article Numbering Code See *bar code*.

European Computer Manufacturers' Association (ECMA) A trade association formed by the major computer manufacturers in Europe, which acts as a pressure group on standards for computers and for their interconnection.

even parity Where a parity bit is set in a data field so that the sum of its bits is even.

evolutionary refinement See *stepwise refinement*.

exception reporting Reporting based on the recognition of specified exception conditions, such as forward orders falling below a given threshold. The computer system checks for these conditions, and only generates a report – in the form of a listing or a message sent to a terminal – when they occur.

exchange Equipment that distributes telecommunications services within a particular geographical area, such as by switching telephone calls.
    See also *packet-switching exchange*; *private automatic branch exchange*.

exchangeable disk A disk storage device from which the storage medium – a disk pack or cartridge – can be removed and stored offline. Compare with *fixed disk*, where the storage medium is permanently sealed within the case containing the drive.

exclusive OR circuit A circuit with two or more inputs and one output whose output is high if any one or more of the inputs (but not all) are low.
    See also *logic element*.

exclusive OR operation A program instruction or statement that performs a boolean exclusive OR operation on two data fields. As shown below, a one bit is set in the output field where one (but not both) of the bits is one in the input fields, otherwise (both ones or both zeros) the bit is set to zero.

Input		Output
0	0	0
0	1	1
1	0	1
1	1	0

execute Perform or carry out. Used of a program executed by a computer system, and of an instruction executed by a processor.

execute cycle The time during which a processor carries out a program instruction. It is preceded by the fetch cycle, when the operands are retrieved from memory ready for processing.

execution time The total time needed by a processor to carry out a program instruction. This is supervised by the control unit, which normally operates in two cycles. It begins by fetching the instruction from memory, then decides what operation is to be performed, identifies the operands and, if necessary, fetches those also from memory. This is known as the fetch cycle. It then enters the execute cycle, activating the arithmetic/logic unit to carry out the processing, then returning the results to memory. Together, the fetch and execute cycles are known as the instruction cycle, and the processor's execution time is the average time needed to complete it – a very rough indicator of throughput potential.

executive information system (EIS) A computer-based system designed to support the work of business executives, by providing information to support decision-making. Typically, an executive information system consists of terminals or personal computers connected to a shared computer that draws data from the organisation's data processing systems and from other sources. Systems are designed for ease of use, for example controlled via a touch screen or a mouse, and present a variety of information in easily assimilated graphical form. EIS emphasise data retrieval and display, by contrast with *decision support systems* (DSS), normally used by lower-level managers, that emphasise decision modelling.

executive program See *operating system*.

EXIT statement A statement used in some high-level languages to leave a loop before the normal termination condition arises.

expanded memory Memory on an IBM PC or a compatible personal computer above the normal limit of 640 kilobytes.

expansion card A printed circuit board that can be plugged into a slot in the mother-board of a small computer such as a personal computer, to extend its capabilities. Machines with expansion slots are referred to as 'open' machines, because manufacturers other than the originator can easily build equipment to work with it by designing an appropriate expansion card, the interface between the expansion slot and the rest of the computer being a published standard. Expansion cards are commonly used to interface non-standard peripherals; to enhance basic capabilities, such as providing high-resolution graphics or a faster processor (see *accelerator card*); and to emulate other computers or operating systems.

expansion slot A slot in the motherboard of a small computer such as a personal computer containing a row of connectors. It is designed to accept a plug-in printed circuit board – see *expansion card*.

expert system A computer system that helps people to find the solution to a problem, drawing on expert knowledge that has been embodied within it (explicitly) in the form of a knowledge base. Usually, the expert system can also explain to the user the line of reasoning it has followed (see *explanation*). An expert system has four main components:

(1) The knowledge base contains the facts agreed by experts, the common knowledge they have acquired over years of work, and the rules of thumb that they apply to derive conclusions.

(2) The inference engine is a program that simulates the deductive thought processes of the human expert, drawing on the knowledge base and further information entered by the expert system user.

(3) The user interface helps the expert system user to enter relevant facts and to ask questions. Normally, it will use a natural 'English-like' language to conduct a question-and-answer dialogue with the user.

(4) The knowledge acquisition facility acquires knowledge from the human expert (or the knowledge engineer who acts as an intermediary between the human expert and the expert system) in the form of rules and facts, and encodes these to build up the knowledge base.

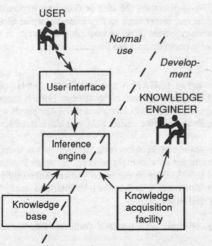

Expert systems fall into two broad classes. The first class of system helps people who are not experts to solve well-defined problems, such as for example to diagnose and perhaps fix faults in a car engine. The second class is used by people who are experts themselves, and who need help to 'navigate' through a very complex domain of knowledge, such as medical specialists attempting to diagnose obscure conditions; engineers attempting to work out why some very complex equipment is not working correctly; or geologists trying to decide where to drill for oil.

Most current expert systems only incorporate what is referred to as shallow knowledge or heuristics. So-called second-generation expert systems also incorporate more general ('deep') knowledge, on which they can fall back when faced with problems that cannot be solved by shallow knowledge alone.

expert system shell A generalised expert system containing no knowledge. It provides a framework for a developer to build a new knowledge base for a particular expert system application.

explanation The ability of an expert system to explain the reasoning process it is using. If the user asks 'why?' in response to a question posed by the expert system, the expert system explains why the information it has asked for may lead to a given conclusion. If the user asks 'how?' after being presented with a conclusion, the expert system explains what rules in the knowledge base it has used to reach it. Some expert systems also respond to variants on these questions such as 'why not?' and 'how could?'.

export Transfer (usually data or documents) out of one application or computer system for use by another. By implication, the data or document exported will be in a format intelligible to the receiver, rather than in any internal format used by the sender. For example, a database management package may allow a data file to be exported so that it can be read by a word processing package for mailmerge.

Compare *import*.

extended graphics array (XGA) A standard for the display of information on the screen of a personal computer, announced by IBM in 1991. It supports two screen resolutions: 1024 x 768 pixels with 256 colours, and 640 x 480 pixels with 65,536 colours. It also provides compatibility with the earlier *VGA* graphics standard.

extended industry-standard architecture (EISA) A bus specification for personal computers, being promoted by a number of suppliers as an alternative to IBM's micro channel architecture (MCA). It is an enhanced version of the original *industry-standard architecture* (ISA), carrying 32 rather than 16 bits, and accepting expansion cards designed for ISA as well as for itself.

extended memory Memory on a personal computer that lies above the 1 megabyte mark.

external memory See *mass storage*.

external schema A schema representing a 'view' of a database as required by a particular applications program or database user. In database terminology, a schema is a programmed description of data items (sometimes described as entities) and of the relationships between them. The external schema (sometimes called a sub-schema) is a subset of the conceptual schema, which represents all of the data held in a database.

Compare *internal schema*.

F

facilities management A service whereby a contractor takes complete responsibility for running an organisation's computer-based systems, sometimes including developing and maintaining the applications programs and/or handling end-users' problems.

See also *outsourcing*.

facsimile transmission (fax) Electronic transmission of copies of documents for reproduction at a remote site. The equipment used for facsimile transmission falls into a number of groups, based on standards defined by CCITT. Groups 1 to 3 can operate over public telephone networks or over private networks:

Group 1: analog transmission at a speed of 4-6 minutes per page;

Group 2: analog transmission at a speed of 2-3 minutes per page;

Group 3: digital transmission at less than one minute per page, with a resolution up to 200 dots per inch, and providing for compression of data.

Group 4 provides for high-speed digital transmission over digital data networks only.

fact template A way of expressing facts in a graphical rule language. The fact template shows the attribute, the predicate and the value of a fact, and the context in which it belongs.

fail softly Where a computer system is able to close down its activities progressively and in a controlled way when failures occur, rather than terminating abruptly. This makes it possible to minimise loss of data, and perhaps to continue service at a degraded level until repairs are carried out. Also known as *graceful degradation*.

failure rate See *mean time between failures*.

fallback Revert to alternative ways of working, such as manual processing, when a computer system fails. Normally fallback working will be designed to keep operations going on a limited basis until the computer can be restored to normal service.

fanfold stationery See *continuous stationery*.

FAST Abbreviation of *Federation against Software Theft*.

fast circuit switching A form of circuit switching designed for computer devices, in which calls can be set up much more rapidly – in fractions of a second – than on the telephone network. Fast circuit switching is economic for applications where large numbers of terminals need to make occasional short calls to one or more computer systems, as in credit clearing.

fatal error A software malfunction, in an operating system or an applications program, from which no recovery is possible. Hence the computer (in the former case) or the program (in the latter) stops abruptly at the point where the error occurred.

father tape See *grandfather-father-son*.

fault time See *down time*.

fault tolerant Able to operate correctly even when faults or failures occur in components. At component level, this is done by including redundant chips and circuits within a component and by building logic into it so that faults are bypassed automatically. At computer system level, it is done by replicating components such as processors and disk drives.

fax Abbreviation of *facsimile transmission*.

FCC Abbreviation of *Federal Communications Commission*.

FCS Abbreviation of *frame check sequence*.

FDDI Abbreviation of *fibre distributed data interface*.

FDM Abbreviation of *frequency division multiplex*.

FDX Abbreviation of *full-duplex*.

feasibility study A study designed to determine whether it is a practical and economic proposition to undertake a project. A feasibility study is appropriate when there are grounds for doubt about the level of benefits that may be realised or about the practicality of possible solutions to the problem under study – such as whether a solution can be developed within an acceptable timescale and budget; whether suitable equipment and/or skills are available; or indeed whether information technology can contribute to solving the problem at all. Also known as *requirements analysis*.

feature extraction A technique used in optical character recognition. The significant features of characters that give them an unique identity are recorded, such as number and type of lines. To identify an unknown character its features are compared with those of the recorded characters until a match is found. This technique is better able to identify characters regardless of typeface or point size variation than the simpler technique called *matrix matching*.

Federal Communications Commission (FCC) The regulatory body for telecommunications in the US.

federal information processing standard (FIPS) A set of standards and guidelines drawn up by the US National Institute of Standards and Technology (formerly the National Bureau of Standards) for use within the federal sector. It recommends standards for open systems interconnection (OSI), and includes an application portability profile (APP).

Federation against Software Theft (FAST) An organisation set up by a number of computer and software manufacturers to prevent and detect the illegal copying of software, and particularly of personal computer software packages. It campaigns for better legislative protection for copyright in software products, and initiates through the courts raids of companies suspected of using illegal copies.

feed holes See *sprocket holes*.

fetch cycle The time in which a processor prepares to carry out a program instruction. In the fetch cycle, the instruction is decoded and the operands are retrieved from memory ready for processing. It is followed by the execute cycle, in which the actual processing is carried out.

fiber optics See *fibre optics*.

fibre distributed data interface (FDDI) A standard for connection to networks using fibre optic cables. It is in the process of definition by ISO. It defines a transmission network operating at a data rate of 100 megabits/second, over a 100 kilometer area, with up to 1,000 connections. (Contrast this with conventional local area network technology such as *Ethernet*, operating at 10 megabits/second up to about 5 kilometers.) Normally the network uses an arrangement of dual, contra-rotating rings – a primary and a secondary. Should a break occur, the network can reconfigure itself automatically by connecting the primary to the secondary to form a new ring. This enormous bandwidth is subdivided into bands, each of which can carry a separate transmission service, using different protocols. An FDDI network can serve as a backbone network connecting a number of local area networks in a campus, a large site or a metropolitan area, or it may connect devices requiring very high bandwidth connections, such as CAD workstations.

fibre optics A transmission technology in which light pulses are transmitted along specially manufactured optical fibres. Each fibre consists of a central core surrounded by a sheath with a much lower refractive index. Light signals introduced at one end of a fibre are conducted along the core because the signals are reflected from the sheath. The light signals may be generated either by light emitting diode (LED) or by laser.

The advantages of fibre optics over copper cable are lower weight and bulk, flexibility, immunity to electrical interference, and better signal transmitting properties. Whereas copper cables require repeaters every few miles to maintain a signal of acceptable strength, fibre optic cables can be laid in continuous lengths of well over 100 kilometres. They also have a far greater transmission capacity than copper cables. The technology is still under development, but already the number of simultaneous telephone conversations that can be transmitted down a single fibre has increased from an original 7,500 to 75,000. However, fibre optic cables are more difficult to join than copper cables, are more expensive per metre run, and also require extra equipment at each end of a line to generate and receive the light signals. This makes them most economic over longer distances, and where their huge transmission capacity is needed.

field See *data field*.

field separator See *delimiter*.

FIFO Abbreviation of *first-in first-out*.

fifth generation computer A term coined by the Japanese to describe a planned new generation of computers based on artificial intelligence principles. They launched the project to develop a fifth generation computer in 1981, intending to produce commercially viable products by the mid-1990s. The project combines research on very large scale integration, parallel processing, pattern recognition, logic programming and knowledge-based systems. It provoked a UK response in the form of the Alvey programme and a European response in the form of the ESPRIT programme.

file See *data file*.

file compression See *data compression*.

file dump A complete record of the contents of data files at a particular point in time, such as at the end of a day's processing. The dump will either be made on a different storage medium, such as by dumping disk files on to tape, or may be a separate copy on the same medium which is stored offline rather than being used for processing. Each dump will normally be kept for a certain period of time so that it can be used as the basis for file recovery if needed.

See also *incremental dump*.

117

file extension A short descriptive code (three characters on most personal computers) added after a filename to identify the format or purpose of the file and to distinguish it from other versions of that same file. Thus, for example, the source code version of a program might be MYPROG and the compiled version MYPROG.COM.

file extent An area occupied by a file on a direct access storage device, consisting of a number of contiguous tracks. One file may consist of a number of file extents, which are linked together by the operating system so that the file appears to programs using it to be continuous.

file label See *header label*.

file locking A means of controlling access to shared files. Programs specify when they open a file whether they require exclusive access to the file, or whether other programs must be prevented from updating the file while it remains open. Subsequently, the operating software refuses requests from (locks out) other programs which try to open the file in any conflicting way. File locking is a simple and effective way of preventing programs from overwriting one another's updates, but it can seriously affect performance where several programs compete regularly for the same files. In these circumstances, locking is applied to a part of the file only, such as a single record or a page.

Compare *record locking*.

file management Processing performed by software, usually contained within an operating system, which helps users or programs to create and manage files held on mass storage such as tape or disk. Sometimes also used to mean management of simple 'flat' files, in contrast to *database management* which handles sets of interrelated files.

file organisation See *file structure*.

file recovery The reinstatement of data files that have been damaged by hardware failure or corrupted by incorrect processing. File recovery is normally carried out by reloading a file dump taken earlier, then rerunning programs or reprocessing transactions that have updated the files since the dump was taken.

See also *transaction journal*.

file server See *server*.

file sharing Where more than one user may access and update a file, accessible via a network, at the same time.

See also *client-server; server*.

file structure The way records are distributed within a data file, including structures such as indexes that are created to help to find particular records rapidly. Records in data

118

files are normally held in blocks of a fixed maximum length, and the structure adopted for each file determines how the file management software decides in which block to place a new record and, subsequently, in which block to find it again. With serial storage media such as magnetic tape, the software has no choice but to write records and blocks serially, as they arise, and records can only be read back in the same sequence. This can be an extremely wasteful way to find a particular record. On direct access storage media such as disks, however, more options are available. Apart from serial, the main structures used for direct access files are:

(1) Indexed sequential: records are held in sequence according to a specified key field. Additionally, an index is created by the software, consisting of key values of records in the file (the key is a specified field within each record), together with the location of the block in which the record may be found. To find a particular record, the software searches the index, then reads the specified block and searches it to find the record. Indexed sequential organisation has the advantage that records can be read in key sequence for batch processing and can also be retrieved directly for transaction processing or enquiry applications.

(2) Random (sometimes known as hashed random or just hash): the block in which a record is placed is determined by applying an algorithm to the key value to generate a pseudo-random number within the range of blocks available for storage. Records in random files can be accessed rapidly and efficiently, but efficiency declines in proportion to the number of records in overflow blocks, and they cannot easily be read in key sequence.

Indexed sequential and random file structures are useful where applications require to access individual records, based on key values, but many applications need to access records based on logical relationships (for example, 'find me all the records of customers who took delivery of product y last month') or based on a combination of key fields (for example, 'find me all the abstracts which mention "Brazil" and "rain forests" or "logging"'). For applications such as the former, a database management system may be used, which uses indexed sequential or random file structures internally to express the relationships within the database. For the latter, which is typical of text retrieval applications, inverted file structures are often used, in which all the fields (or words for text files) in each record are recorded in an index.

file transfer The transfer of a file – a data file, a complete document, an object program – from one computer system to another, so that it can be processed on the receiving system.

file transfer access and management (FTAM) A protocol for file transfer and for remote file access. It allows a user application to transfer files across a network, to read or write files on a computer system located elsewhere in a network, and also to create or delete such files or to change their attributes. The protocol has been approved by ISO as part of the open systems interconnection standards (coded ISO 8571).

file update See *record update*.

filestore See *mass storage*.

filter bridge See *bridge*.

finance lease A conventional leasing arrangement that gives the lessee the benefits of full ownership of equipment without having to pay the total purchase price up front. For large computers, operating leases offering greater flexibility are more popular.

financial modelling Use of a computer to simulate and predict the financial implications of alternative courses of action or circumstances.

finite element analysis The use of mathematical techniques to predict the behaviour of an entire manufactured part, based on the behaviour of its discrete mathematical sub-elements.

FIPS Abbreviation of *federal information processing standard*.

firmware Software built into integrated circuits in a permanent or semi-permanent form. Firmware lies midway between hardware and software in terms of performance and flexibility. It is less flexible than software, that can be modified simply by reading a new version into memory, but has a better performance since the program instructions are built into an integrated circuit and therefore do not need to be retrieved from memory before they can be executed. Firmware is used to hold sections of operating system code or input/output routines that are used frequently but change rarely, where the better performance of firmware justifies its extra cost. Firmware is held in semiconductor memory, such as read-only memory (ROM) and more flexible variants like EPROM and EAROM.

first generation language See *machine language*.

first-in first-out (FIFO) Describes how items in a queue or list are processed. As with a bus queue, the first arrivals in the queue are the first to leave it. Contrast with *last-in first-out* which is the method used for a push-down stack.

fixed disk A disk storage device from which the storage medium – the disk pack – cannot be removed, but is permanently sealed within the case containing the drive. Contrast with *exchangeable disk*, where the storage medium can be removed and stored offline. The most widely used type of fixed disk is the *Winchester disk*, used on many personal computers.

fixed-length record A record whose length remains the same throughout a file. The

length may be specified in words, bytes or characters.
Compare *variable-length record*.

fixed point arithmetic Arithmetic calculations that treat operands and result as integers, without changing the relative position within them of a decimal point.
Compare *floating point arithmetic*.

flag (1) In data communications, a fixed bit pattern (consisting of 01111110) marking the beginning and end of a frame using bit-synchronous communications procedures, such as in packet switching.

(2) In programming, a variable that is set by the program as a reminder that a condition has occurred. When the program reaches a convenient point, the flag is tested and appropriate action is taken. Also known as *indicator*.

flash chip Computer memory consisting of a particular type of EEPROM (electrically erasable programmable read only memory) that can be erased by a pulse of electricity rather than by ultraviolet light. It is expected to be competive with hard disks for data storage in portable computers by the mid-1990s. It has no moving parts, so should be more reliable, and since it is nonvolatile requires no battery backup.

flash EEPROM See *flash chip*.

flat file A file consisting of only one type of record, with no physical relationship with other files. In other words, effectively a two-dimensional table, one dimension consisting of the records and the other of the fields within each record. Flat files are simple to construct and process, but cannot represent directly the complex relationships between data items that frequently occur in real life.
Compare *database*.

flexible manufacturing system (FMS) A computer-controlled manufacturing system that can produce families of parts, usually in small batches, without expensive manual changeovers. It consists of groups of computer numerically controlled (CNC) machine tools linked by automatic conveyors.

flicker An important measure of the quality of a visual display screen. Flicker can contribute significantly to the tiredness and eyestrain experienced by people working with visual displays. At a limited cost, screens can now be made virtually flicker-free.

floating point arithmetic Arithmetic calculations based on floating point numbers. Instead of being stored in normal decimal form, floating point numbers are represented by a fixed point part and an exponent, which means that the relative position of the decimal point need not be constant but 'floats'. Thus 347.45 would be stored as 3.4745×10^2, and 3474.5 as 3.4745×10^3. Floating point arithmetic is used in computer systems for complex calculations, for very large numbers, and where a high level of accuracy is needed.

Compare *fixed point arithmetic*.

floating point unit (FPU) An additional processor designed expressly to perform floating point calculations, sometimes referred to as a maths coprocessor. It can be added to a computer to speed up its performance for scientific or mathematical applications.

floppy disk A disk storage medium made of flexible material. The original floppy disks were made of plastic with magnetisable particles on its surface, supplied sealed in a plastic envelope with holes cut in it for the drive shaft and for the recording heads, to protect the recording surface. Standard sizes were 8.25, and later 5.25 inches in diameter. They are gradually being superceded by 3.5 inch disks in a harder, but still slightly flexible plastic case, with a sliding metal cover protecting the hole for the recording heads. These disks are capable of holding more data than their larger precursors – up to about 1.5 megabytes on a single disk. Also known as *diskette*.

FLOPS Abbreviation of floating point instructions per second, a measure of the performance of a processor with a scientific or mathematical workload.

Compare *mips*.

flow control Control of the rate at which information is exchanged over a transmission line or a communications link. Flow control is necessary whenever one of the devices engaged in an exchange cannot always accept information at the rate that the other device can deliver it. Normally this is because the receiving device needs time to process one unit of information before being able to accept another. Sometimes the receiving device uses buffers to hold a 'stock' of information and thus even the flow, in which case flow control is geared to the availability of space in the buffer.

flowchart A diagram representing a sequence of events or operations. Flowcharts are used to depict a range of activities, including clerical processes, computer applications and programs, and at varying levels of detail. The most commonly used flowcharting symbols (see figure opposite) are rectangles containing descriptive text, used for processing operations, and diamonds, used to represent decisions or alternative courses of action.

flyback period See *vertical blanking interval*.

FMS Abbreviation of *flexible manufacturing system*.

font A complete set of characters – letters, numbers, punctuation marks, special symbols – designed to have a consistent appearance and belonging to one typeface. Fonts are identified by names such as Times and Helvetica.

Flowcharting symbols

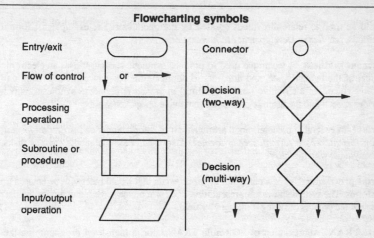

Entry/exit	Connector
Flow of control	Decision (two-way)
Processing operation	
Subroutine or procedure	Decision (multi-way)
Input/output operation	

font family A font in a range of sizes and styles.

footprint (1) The space that a device such as a personal computer or data terminal occupies on a desk.

(2) The geographical area to which a satellite broadcasting system can transmit directly – see *direct broadcasting by satellite*.

FOR statement A control structure used in some high-level languages to construct a loop within a program. The statement takes the general form 'FOR control expression DO action', indicating that the statements constituting 'action' are to be repeated as specified in 'control expression', normally using a control variable.

foreground processing The highest priority processing within a computer. On a personal computer, the foreground application controls the keyboard and display screen.

Compare *background processing*.

form feed Movement of the paper to the head of the next page, in a printer using continuous stationery. May also be used to cause the next sheet to be fed through a sheet feeder, ready for printing.

form letters See *mailmerge*.

formal method A method of describing systems or programs that is based on mathematical methods, and hence the correctness or otherwise of the specification can be proved. A formal method using algebraic notation, for example, would have a syntax to specify the operation of a system, and semantics consisting of a set of algebraic equations that

123

could be used to relate the values created by the operations. For examples of formal methods, see *Vienna Development Method* and *Z*.

formant synthesis A technique used to produce synthetic speech. It uses an electronic model of the human voice, and attempts to duplicate the resonances of the rising and falling voice. An alternative (and prevailing) approach is phoneme synthesis, which divides speech into the discrete phonemes that make up spoken words.

format In general, a predetermined arrangement of information. See *instruction format*, *print format*, *record format*, *screen format*. Also used of storage media such as disks, synonymous with *initialise*.

forms processing Use of a computer to capture data that would otherwise be entered up on forms. The system shows a representation of a form on the screen, and guides the user through the process of filling it in, field by field.

FORTRAN Abbreviation of FORmula TRANslator, a high-level programming language designed for scientific and mathematical applications. It was introduced in 1974. FORTRAN is a simple language, written using a combination of algebraic formulae and English-like statements. In its original version even simple file processing was quite difficult to achieve, but it has since been extended both for file processing and for graphics.

forward chaining A control strategy used by the reasoning mechanism of an expert system (the inference engine). The program starts by satisfying a set of conditions, then works forward towards some (possibly remote) conclusion.
　　Compare *backward chaining*.

forward reasoning See *forward chaining*.

fount See *font*.

4GL Abbreviation of *fourth generation language*.

four-wire circuit A transmission path that provides two separate channels for information flow. Normally, one channel is used to transmit the information and the other for control signals. Four-wire circuits are used for special applications, such as with some types of electronic telephone exchange, and for long-distance multiplexed links. Normal telephone wiring consists of two-wire circuits.

fourth generation language (4GL) An advanced programming language. Originally the term was applied to powerful languages or toolkits associated with database management systems, that were used by specialist programmers to generate applications programs that used the database. It is often used in a looser sense to mean any powerful system building tool, including those designed for end-users rather than specialist programmers. Programmers can be more productive with fourth generation languages than with earlier languages such as COBOL because fewer statements are needed to achieve the same results.

The earlier generations of programming languages are:

(1) *machine language* programming;

(2) *assembly languages*;

(3) *high-level languages* like COBOL and FORTRAN.

Compare *application generator*.

FPA Abbreviation of *function point analysis*.

FPU Abbreviation of *floating point unit*.

frame (1) In videotex and teletext systems, a screenful of information.

(2) In data communications, a unit of information, bounded by framing characters – see *bit-synchronous protocol*.

(3) In video applications such as video conferencing or interactive video, a single still picture in a moving picture sequence.

(4) In expert systems, a way of representing knowledge. Each frame represents a stereotyped situation, and shows both descriptive data about objects or ideas and procedures related to them.

frame check sequence (FCS) A field containing a longitudinal redundancy check – see *bit-synchronous*.

frame relay A data network interface specification designed for high-speed transmission of data, such as to send very large image files or to interconnect local area networks. It combines features of both packet switching and circuit switching (variable delay along with high throughput) at the cost of limited error detection and correction. It works best over lines of around 2 megabits/second or faster.

freeze frame video A form of video transmission in which the video picture changes every few seconds or every few minutes, rather than continuously as in a normal television picture. Also known as *slow scan video* or *stop action video*. Freeze frame video is cheaper to install and transmit than full motion video, and can operate over ordinary telephone lines. It is used where it is not necessary to have a continuous record of activity, such as to show slides or pictures accompanying a verbal explanation, or for security monitoring.

Compare *full motion video*.

frequency division multiplex (FDM) Share a transmission channel by dividing its

125

bandwidth into several parallel paths, defined by bands of wave frequencies. This is like the technique used to ensure that FM radio stations do not interfere with one another, by assigning each a separate transmission frequency.

Compare *time division multiplex*.

frequency shift keying (FSK) A method of transmitting binary information. The frequency of a carrier signal is varied between two levels, without changing the phase of the signal, to represent binary 0 or 1.

friction feed Paper fed through a printer by pressure of rollers against the paper, rather than by pins in sprocket holes (pin or sprocket feed).

front-end processor A processor co-located with the main processor of a large computer system, and attached between it and the communications network. The front-end processor is designed specifically to handle the messages flowing to and from the computer system via the network, relieving the main processor of tasks which it cannot handle efficiently.

Compare *back-end processor*.

FSK Abbreviation of *frequency shift keying*.

FTAM Abbreviation of *file transfer access and management*.

FTP See *TCP/IP*.

full-duplex (FDX) Of a transmission link where transmission can take place in both directions simultaneously. Contrast with *half-duplex*, where transmission can take place in either direction, but only in one direction at a time.

full motion video A form of video transmission in which the video picture changes continuously as in a normal television picture. Contrast with *freeze frame video* (also known as *slow scan video* or *stop action video*) where the picture only changes every few seconds or every few minutes. Full motion video is used where it is important to have a continuous record of activity, such as for videoconferencing.

full text retrieval Retrieval of text from a file (of documents or abstracts) using any combination of words (or part words) contained in the text. Full text retrieval systems often use inverted file structures and can be contrasted with keyword retrieval systems, where only a specified range of keywords can be used to identify the records that are required. A user at a terminal normally begins a retrieval operation by finding out how many records match an initial set of search criteria. If necessary, the next step is to narrow the search by adding further criteria. Finally, when a small enough number of records that meet the criteria have been identified, all those records are retrieved and can

be examined on the display screen.
See also *associative retrieval*.

full text search See *full text retrieval*.

function A procedure in a high-level language that returns a value. It may either be a standard function built into the language or a function written by the programmer for his or her own purposes. A function is normally called simply by including its identifier (with appropriate arguments) in a program statement. The PASCAL statement 'a := SQRT(x)', for example, puts the value of the square root of variable 'x' into variable 'a'.

function key A key, usually one of a row above the normal typewriter keyboard, that is used to activate special functions, rather than to enter or format text. Some function keys may be used for a similar purpose in all applications, such as to call in help routines. The effect of others will vary depending on the application that is running.

function point analysis (FPA) A way of measuring productivity in developing software. 'Function points' were invented by Dr Alan Albrecht of IBM as a measure of the size of a computer system, and are counted by the process known as function point analysis. It considers only those aspects of a system that are apparent to the user, rather the complexity of the system itself, which may only be a reflection of the methods adopted to develop it. Function points per person day can then be used as a measure of productivity. Function point analysis is seen as better than other measures of productivity for two reasons – firstly because it focuses on what the user gets rather than what the programmer does, thus taking account of activities such as requirements analysis and systems design that precede programming, but may well be more important; and secondly because function points are independent of the equipment or the programming language that is used, or how the development team is organised.

functional decomposition Breaking down of a business process or a computer program into the operations it comprises. A technique used in the design of applications and programs.

functional profile A specified subset of interconnection standards, designed to meet the requirements of a particular class of user. The reason for specifying a profile, also known as a *functional standard*, is to restrict the wide range of choice available in the base standards, and thus simplify the task of introducing them. See, for example, GOSIP (*Government OSI profile*) and *application portability profile*.

functional specification A document explaining WHAT a computer application (or, more rarely, a program) is to do, rather than HOW it is to do it. A functional specification is normally produced by a systems analyst in the early stages of a large development project, written in a language that the intended user can understand. Often, the user must formally approve the functional specification before further design or development work can start. Sometimes known as requirements specification or operational requirement.

See also *development lifecycle*; *system documentation*.

Compare *system specification*.

functional standard See *functional profile*.

functionality The measure of how many different purposes a piece of hardware or software can serve.

fuzzy logic A method for handling inexact information by attempting to quantify value judgements. It incorporates rules for manipulating fuzzy sets, which are sets of values corresponding to a fuzzy proposition. For example, the fuzzy proposition 'Fred is a tall man' might correspond to the fuzzy set:

- the probability of Fred being less than 1.7m is .1;
- the probability of Fred being between 1.7m and 1.9m is .7;
- the probability of Fred being over 1.9m is .2.

See also *expert system*; *logic programming*.

G

---

gallium arsenide A material used for integrated circuits (chips) as an alternative to silicon. It has greater conductivity than silicon and therefore, in theory, should work up to five times faster and use much less power. These theoretical advantages have proved difficult to realise in practice, however, and the manufacturing cost of gallium arsenide chips is much higher than for silicon chips, mainly because of low yield (in other words, the proportion of chips in a manufactured batch that are useable).

gamut See *colour gamut*.

gas plasma See *plasma display*.

gate See *logic element*.

gateway A piece of equipment that connects a data communications network at one location to outside services and networks, and to networks at different locations. A gateway enables devices attached to its own network to communicate across other networks, such as public packet-switching networks, and with other computers, such as the corporate data centre, that do not use the same communications procedures as they do. In terms of the OSI reference model, gateways operate up to levels 6 or 7, the presentation and application layers.

Compare *bridge*; *repeater*; *router*.

GCR Abbreviation of *group-coded recording*.

general-purpose computer A computer system capable of running a varied workload efficiently, normally including batch processing, teleprocessing and scientific computing. Special-purpose computers, by contrast, would be designed to cope with only one of those or perhaps only with a single application, such as aircraft simulation.

generalisation hierarchy A database concept which allows the database to model hierarchical relationships. It does this by allowing one subclass to exclude another. For example, if there is a class of data called 'people', which contains a further breakdown into 'men' and 'women', these sub-classes can be defined as excluding one another. In other words, 'men' cannot be 'women' and vice versa.

generation A major stage in the development of computer hardware or software. See, for example, *fifth generation computer* and *fourth generation language* (4GL).

generation number A number, associated with a mass storage file, which identifies its age or generation. It is used with serial files, such as on tape, where a complete new version of the file is produced every time it is updated, using a batch of amendments. The name of the file remains the same from one updating run to the next, but the generation number increases by one. This makes it possible to check that the latest generation of a file is being used, and also makes it possible to recover when the latest generation of a file is damaged, by processing the appropriate batch(es) of amendments against an earlier generation.

generic computing Computing tools designed to support a generic class of activities, rather than a specific business function. Spreadsheet and word processing are examples of generic computing tools, whereas payroll and stock control, for example, are specific applications.

genlock In interactive video applications, the writing of computer-generated information in step with, and in the same electrical form as, an external video signal. It is applied where images from two sources are to be combined, such as by superimposing or by dissolving from one to the other.

geographic information system (GIS) An information system that incorporates maps in digital form. They are used, for example, by public utilities to record the location of distribution equipment and to plan maintenance and repair work.

geometric graphics See *object graphics*.

ghosting See *shadowing*.

Gibson's mix A mix of instructions designed to provide a measure of the average instruction time of a processor with a typical workload, widely used during the 1960s. In other words, a more accurate measure of instruction time than a raw mips (millions of instructions per second) figure.

giga- A prefix meaning thousands of millions of (x 10^9).

GIGO Abbreviation of *garbage in, garbage out*, meaning that unreliable input to a system will produce unreliable results.

GIS Abbreviation of *geographic information system*.

GKS Abbreviation of *graphics kernel system*.

glare A problem frequently experienced by people working in front of visual display screens. Glare comes directly from the display, rather like driving at night with a dirty windscreen. There may also be indirect glare reflected off other surfaces surrounding the screen. Glare can be reduced by good lighting and, where the display itself is at fault, by fitting an antiglare filter over the screen.

glare filter See *antiglare filter*.

glass teletype A visual display terminal that operates in exactly the same way as a teletype keyboard terminal, except that text is displayed on the screen rather than being printed out.

global variable A variable that is accessible to all the elements of a program, rather than just in the element in which it is declared.
    Compare *local variable*.

GO TO statement A statement used in some high-level languages to branch unconditionally to another point in the program. It has the general form 'GO TO identifier', where 'identifier' is the name assigned to a statement. The GO TO statement (or rather excessive use of it by programmers) has been criticised as contributing to badly structured programs that are difficult to maintain. So-called block-structured languages are designed so that GO TO statements are unnecessary.

GOSIP Acronym for *government OSI profile*, a subset of open systems interconnection (OSI) standards promoted, in the UK, by the Central Computing & Telecommunications Agency (CCTA) for government computing. The profile specifies a number of different types of data network, and a limited range of application level services that any of them may offer. Similar profiles have been defined in other countries for public sector computing.

graceful degradation See *fail softly*.

grandfather-father-son A technique used to secure data files processed in batch, particularly on magnetic tape, against accidental damage or loss. Whenever a file is updated, a completely new version is produced on a fresh tape. Three versions of the tape are kept – known as grandfather, father and son, and identified by the generation number

– along with the amendment tapes. If the 'son' copy is damaged or lost, it can be recreated by running the amendments against the 'father'. The 'father' can be recreated in a similar way if it too is unusable.

grandfather tape See *grandfather-father-son*.

graph plotter See *plotter*.

graphic character See *printable character*.

graphical browser A diagrammatic overview of the items of information (the nodes) and the links between them in a hypertext database. It helps the user to find his or her way around within the database, a process known as browsing.

graphical rule language A language used to express the rules contained in a knowledge base. It consists of rules structured by means of a rule template.

graphical user interface (gui) A general term for a user interface (the mechanisms that enable people to converse with computers) based on graphics rather than solely on keyed text. Such an interface was developed originally by Xerox and achieved commercial success and popularity in the form of Apple's Macintosh personal computer. It is now being introduced to IBM-compatible personal computers through Microsoft's Windows software.

    Also known by the acronym WIMPs, for Windows, Icons, Mouse, Pull-down menus – the four key elements of these interfaces.

    Compare *command line interface*.

graphics See *computer graphics*.

graphics kernel system (GKS) A standard defining how applications programs should represent and display two-dimensional graphics that is independent of any particular programming language. Applications programs that conform to the standard can exchange graphical information directly, rather than having to print it out on paper to transfer to others. It has been approved by ISO as part of open systems interconnection standards (coded ISO 7492).

    Compare *initial graphics exchange specification*; *PHIGS*.

graphics tablet A device used to input graphics to a computer system by hand. The tablet has a flat surface in which a grid of wires has been embedded. The user draws or indicates positions on the tablet with a stylus which makes electrical contact with the wires. The tablet sends the coordinates of the stylus position to the computer system, which responds, as appropriate, by moving the cursor on the screen, drawing lines, etc.

    See also *pen computer*.

graphics workstation See *workstation*.

grey importer A distributor of computer equipment or software who obtains supplies from an overseas dealer, usually in the country of manufacture, rather than operating as an authorised dealer and obtaining supplies via the manufacturer's distribution network in its own country. 'Grey' importing occurs because manufacturers sell to dealers at higher prices in some countries than in others, claiming as justification higher costs for doing business. For example, US equipment is sometimes sold in the UK at an effective exchange rate of 1$ to the £.

grey linearity A measure of the quality of a grey-scale or colour display, reflecting how evenly it can produce each shade of grey. Ideally, each should contain equal amounts of red, green and blue.

grey scale Able to distinguish shades of black, rather than black and white alone. Used of scanned images and the visual display screens and printers which reproduce them. Grey scale images give a more convincing representation of high-quality photographs than monochrome (black and white only) images, but require about three times as much storage space.

group 3 facsimile Standards for the transmission of information between facsimile transceivers, agreed by CCITT in late 1980. They effectively made obsolete the group 1 and 2 standards in operation previously, which were for analog transmission at speeds of 4-6 minutes and 2-3 minutes per A4 page, respectively. Group 3 standards specify digital transmission at less than one minute per page, with a resolution up to 200 dots per inch, and providing for compression of data.

group-coded recording (GCR) The standard used by Apple personal computers to initialise disks.
Compare *modified frequency modulation*.

groupware A term coined to describe software tools designed to help small teams of people work together more effectively, in the same way that personal computer software helps the individual. This includes, for example, software packages that help people to cooperate in writing documents.

gui Abbreviation of *graphical user interface*.

H

---

hack Originally meant to program software or alter programming code, usually operating outside the formal business structure of the software industry, but is now associated more with attempting unauthorised access to a computer than with programming. This breed of hackers has gained a bad name because of their much-publicised successes in breaking through the security procedures surrounding Government computers, but hackers have also been the source of a great deal of valuable public domain software for home and personal computers, distributed via bulletin boards and from user to user.

hacker An individual who engages in hacking.

half-duplex (HDX) Of a transmission link where transmission can take place in either direction, but only in one direction at a time. Contrast with *full-duplex*, where transmission can take place in both directions simultaneously.

halftoning Conversion of continuous tones, such as in a photograph, into a series of graduated tones. Traditional printing techniques achieved this by projecting the image through one or more glass screens with a grid that focused light on to different areas of a piece of film below (hence the alternative term for this process: screening). The digital equivalent of this process begins by scanning the image to record the grey value of each pixel. Output devices such as laser printers build up images as a grid of tiny black dots, so the image is then broken up into patterns of dots (spots) that appear the required shade of grey when printed – the larger the spots the darker the shade.
See also *screen frequency*.

hand-held terminal A small device held in the hand, normally consisting of a keyboard or keypad and a small display, perhaps with a wand attached for reading bar codes. These devices are used for tasks such as inventory control in a supermarket or warehouse. Stock quantities and product codes are keyed or scanned into the terminal and recorded in its memory as the user walks round checking stock levels. Afterwards, the terminal is plugged into a unit that reads the stored information and transmits it to a computer system.

handle A pointer to a pointer. Handles are sometimes used by programs to address data areas in computer systems using dynamic memory management. The memory management routines in the operating system can safely move data around and change the pointers accordingly, without running the risk that a program has stored internally a pointer containing an old, and now incorrect, address.

handshaking Preparation for a transfer of data between two devices, in which each establishes that the other is ready for the transfer to take place.

hard copy Output from a computer system that has a permanent, physical form, such as on paper. Contrast with *soft copy*, such as display on a visual display screen.

hard disk A disk storage medium or device on which the recording surfaces are rigid, in contrast to a *floppy disk*, where the recording surface is flexible.

hard sectoring Of floppy disks requiring a number of positioning holes in the disk to enable the drive to locate the sectors on the disk. Contrast with *soft sectoring*, where only one hole is needed.

hard-wired logic Logic that is fixed at the time of manufacture or assembly, such as by soldered connexions on printed circuit boards. Unlike software or firmware, therefore, it cannot readily be changed.

hardware The physical, manufactured components of a system, or in negative terms those parts of it that are not software. Computer system hardware can be broken down into four main elements:
    (1) the processor that executes the programs;
    (2) the memory that stores programs and data that is being processed;
    (3) the peripherals, used to store data longer term and to exchange commands and information with people;
    (4) the input/output logic that interconnects the above.
    In addition, computer systems frequently use telecommunications services, whose hardware components include transmission lines, switching exchanges and multiplexors.

hash total A control total formed by combining the values in a series of data fields in some way. The result has no significance other than as a check on the contents of the fields concerned.

hashed file See *random file*.

hashing algorithm An algorithm used to derive an address within a specified range from a key value. A hashing algorithm is used with a random file to determine the address of the block in which a given record should be stored.

Hayes-compatible Describes modems that conform to a standard set by a leading US manufacturer of low-cost modems. The standard specifies the control messages that a computer can send to the modem, such as to instruct it to dial a number, and how it responds to those messages. Hayes-compatible modems are widely used with personal computers.

HDLC Abbreviation of *high-level data link control*.

HDX Abbreviation of *half-duplex*.

head crash When the read/write head of a disk drive touches the recording surface. As a result data recorded on the disk will be lost and the disk drive itself may be seriously damaged.

head end The operational centre of a cable television service.

header A data record that precedes a set of detail records. It will normally contain data common to the set. In an order record, for example, the header would contain the identity of the customer, delivery date, and so on, and would be followed by detail records for each line in the order.

header label A special block written at the beginning of a tape reel which identifies the information recorded on it. It contains the name of the file, its generation number, reel number, retention period, and the date when the file was first written. This information is checked by the operating system before a tape is used to verify that it is the one that a program requires. The operating system also checks that the retention period has expired before allowing a tape to be over-written.

heap An area of computer memory, space in which can be allocated and released on demand from programs. It is normally used to hold data structures that vary in size while a program is running.
See also *stack*.

help function See *help screen*.

help key See *help screen*.

help screen Instructions or advice on how to use the current application, displayed on the screen when the user asks for help, such as by pressing the help key (a function key

136

reserved for the purpose) or by choosing 'Help' from a menu.

heuristics Informal, judgmental knowledge of an application area – in other words, rules of thumb. Heuristics also encompass the knowledge of how to solve problems effectively; how to improve performance, and so on. This type of knowledge is accumulated by experts in the field concerned and represents their experience. It is incorporated in the knowledge base of an expert system, along with relevant facts about the problem domain (the intended field of application). Also known as *shallow knowledge*.
    Compare *deep knowledge*.

hex Abbreviation of *hexadecimal notation*.

hexadecimal notation (hex) A notation of numbers to the base sixteen. The ten decimal digits 0 to 9 are used, followed by the alphabetic characters A to F to represent from ten up to fifteen. Hexadecimal notation is used to represent the contents of computer memory and storage devices in preference to binary notation, because it is far more compact; and in preference to decimal notation again because it is more compact, and also because it is easier to interpret. This is because each hexadecimal digit represents exactly four bits, so that codes consisting of four, eight, and even six bits can easily be isolated and recognised. Program instructions, for example, may consist of an 8-bit operation code, followed by operands and addresses of 4, 8 or 16 bits.

hidden backlog A term coined to describe those applications that are needed but do not figure in the applications backlog – the queue of applications waiting to be developed by the computer department. The people who need them have not formally requested to have them developed, probably because they think it so unlikely that they will be approved and delivered that it is not worth the effort.

hierarchical model A model for the structure of a database that sees it as a hierarchy. In other words, any record in a database may be related to (and linked logically with) one, and only one, 'parent' record and/or be related to one or more 'child' records. For example, an order record is linked upwards to the record of the customer who placed the order, and downwards to the records representing the individual items in the order. Databases organised according to the hierarchical model tend to be efficient in processing terms, but lack flexibility.
    Compare *network model*; *relational model*.

high-level data link control (HDLC) A protocol for controlling bit-synchronous data communication across a transmission link, usually referred to as HDLC. HDLC is included in the CCITT packet-switching interface standard known as X.25.
    Compare *synchronous data link control*.
    More detail *frame*.

high-level language A programming language in which each statement corresponds to a number of machine language instructions. Contrast with *assembly language* (or *low-level language*), in which each statement corresponds more or less one-for-one with the instructions that the computer processor executes. The most widely used high-level languages are *COBOL* and *PL/1*, used for commercial applications; *ALGOL* and *FORTRAN*, used for scientific programming; and *BASIC*, available on most personal computers.

Compare *fourth generation language*.

high order See *most significant*.

High Sierra A standard for digital recording of information on CD-ROM compact disks. It is an extension of the Yellow Book standard defined by Philips and Sony, and was agreed by a number of manufacturers in a meeting at the High Sierra hotel near Lake Tahoe in 1986. It allows hierarchical filing of data compatible with many different computer systems.

high-speed modem A modem capable of operating at speeds at and above 9600 bits per second.

highlight See *reverse video*.

highway See *bus*.

hinting A technique used to improve the legibility and visual appeal of computer-generated fonts. Software modifies the outline of each character to suit its size and the nature of the output device, which could be a display screen, a laser printer, or a typesetting machine.

hit list A list of the records or other items found after a search. For example, a hit list may be shown on the screen after a query has been entered specifying the criteria to be used to search a text file, such as a file of abstracts from technical journals. The user can then inspect the hit list to decide whether to narrow the search criteria or to display the records in full.

hit rate The proportion of times that accesses to required items of data are successful. Thus the hit rate for a query on a data file would be the proportion of records that matched the search criteria.

Compare *activity rate*.

holding time The duration of a call (such as a telephone conversation or a data processing transaction) across a communications network. On public networks, the average holding time for voice calls (five minutes or so) is generally much longer than the average holding time for data calls (10-20 seconds).

holography The creation of three-dimensional images (not necessarily by computer means).

home area The area in a direct access file where records are stored based on addresses calculated directly from some value contained within the record itself, meaning that they can normally be retrieved in one access to disk. When there is not enough space in the home area, records are stored in an overflow area, and a 'tag' (or pointer) is stored in the home area.

home computer A personal computer designed for use in the home and therefore cheaper and less robust than those designed for business use.

homeworking See *telecommuting*.

homologation See *network attachment*.

honorware See *shareware*.

horizontal parity check See *longitudinal redundancy check*.

horizontal tabulation Movement of the cursor on a display, or the print mechanism of a printer, horizontally from left to right, to preset positions within the line. Electronic equivalent of the TAB key on a typewriter.

host A computer system attached to a communications network, providing a service or services which other computers and terminals can access via the network.

hot link A feature of some integrated software packages that enables the user to incorporate data prepared in one application into another. When the user copies data, such as from a spreadsheet to a business graphics application to create a graph, the software tracks all subsequent changes to either version and ensures that they remain consistent. Usually, changes to one version are instantly reflected in any other versions displayed on the screen.

Compare *publish and subscribe*.

hot spot See *hyperlink*.

housekeeping The activities necessary to keep a computer system, or some aspect of its operations, in good order. For example, housekeeping of magnetic tape files includes writing header labels and serial numbers on new tapes; removing expired tapes from service; and so on.

hub polling A method of polling terminals (see *poll/select*). The controlling device first invites the furthest terminal on a line to send or receive information. When it has finished, that terminal passes on the invitation to the next furthest terminal, and so on until all terminals on the line have had their turn.

Huffmann encoding A data compression method that works by using a table of character or pattern frequencies to convert symbols in the data into shorter codes. The table may be fixed in advance using the known characteristics of the data, as it is for example for facsimile transmission, where the data normally consists of long runs of white pixels interspersed with short runs of black. Alternatively, and this is the method used by a number of today's modems and data compression packages, it can be built up dynamically by examining the frequency of symbols in the data, allocating the shortest codes for those used most frequently.

See also *LZW algorithm.*

Compare *JPEG algorithm.*

human-computer interaction A general term meaning the methods and mechanisms through which people communicate their requirements and their data to computer systems. Also used (as human-computer interface) as a non-sexist alternative to man-machine interface. Normally, it refers to the ways in which people interact directly with computer systems, such as through visual display screens and keyboards. Work on improving methods of human-computer interaction has been proceeding for many years, but the arrival of the personal computer on so many desktops and in so many homes has brought the issue to the fore. Research and development is focusing on two main areas:

(1) the ways in which information and choices available to the user are presented on the screen, so that he or she can more easily 'navigate' through applications. The most significant work in this area began in Xerox Corporation's Palo Alto Research Center in the 1960s, and was first realised in the WIMPs (Windows, Icons, Mouse, Pull-down menus) interface of Xerox workstations and, later, Apple Computer's Lisa and Macintosh personal computers, now commonly referred to as a *graphical user interface.*

(2) the ways in which users express their instructions. Here, the aim is to use artificial intelligence techniques, such as expert systems and natural language processing, to make it easier for non-specialists and non-programmers to express their requirements in terms the computer can understand.

human factors engineering The design and disposition of equipment in accordance with the requirements, limitations and preferences of people, so that people and equipment can interact effectively and safely. In the field of information technology, there has been great emphasis on equipment issues such as the effects of radiation from visual display screens, but in fact human factors engineers (also called ergonomists) are concerned with three main areas:

(1) the equipment and the systems of which it forms a part;

(2) the design of the workplace in which equipment is installed;

(3) psychosocial factors such as job design and training, which help to determine

whether or not equipment is accepted.

hybrid computer A computer in which analog and digital computers are combined.

hyperlink An area on the computer screen, such as a word or part of a graphic, that the user can activate, such as by indicating it with the mouse pointer and clicking, in order to follow a link to an associated unit of information – the essential principle of hypertext.

hypermedia Multimedia (i.e. a combination of moving pictures, sound, text, graphics) applications which also contain hypertext elements. In other words, the user can follow associative links between units of information, such as by clicking on part of a graphic with a mouse.

hypertext A way of managing information that allows items of information to be connected using associative links. This is believed to be a better model of human mental behaviour than a conventional structured database. A hypertext 'database' (not precisely the right word because, as already indicated, hypertext information is organised differently from a conventional database) consists of a number of nodes connected by links. A node is a collection of information that can be accessed via a single screen, including text, graphics, and, sometimes, sounds or moving picture sequences. The links connecting the nodes are also represented on the screen, and can be followed by the user with a simple action, such as clicking the mouse button or hitting a key. This enables the user to 'browse' through the information with considerable freedom. Hypertext systems usually provide other aids to navigation through the information, such as a *graphical browser* – an overview of the nodes and the links between them in diagram form. Hypertext systems can be used both by individuals to organise their own information, and by groups for collaborative work. The most widely used products are *Notecards* from Xerox and *Hypercard* from Apple.

I

i/o Abbreviation of *input/output*.

i/o bus See *bus*.

i/o channel Abbreviation of *input/output channel*.

i/o interrupt Abbreviation of *input/output interrupt*.

IBIS Abbreviation of *issue-based information system*.

IBM-compatible Used to describe equipment, ranging from personal computers to large mainframe processors, that can run operating or applications software written for equivalent IBM computers without alteration. IBM holds a large share of the market in a number of branches of information technology, and as a result is able to set de facto standards for equipment that less powerful manufacturers sometimes choose to adopt, rather than attempt to introduce their own way of doing things.
    See also *compatibility*.
    Compare *plug-compatible*.

IC Abbreviation of *information centre*.

IC Abbreviation of *integrated circuit*.

icon A small symbol or drawing on a display screen that represents a resource or a function available on the system. For example, each application on Apple's Macintosh personal computer uses a particular icon to represent the documents or files it processes, with an identifying name written underneath. Thus the user can see at a glance whether a file is a word processing document, a data file, a graphic, or whatever. Other icons are used to represent disks, folders that can hold a number of files, and the wastebasket, into

which icons are 'dragged' when the files they represent are no longer needed.
See also *graphical user interface*.

ICR Abbreviation of *intelligent character recognition*.

identification division The part of a COBOL program that contains the name and other identifying details of the program.

identifier An identifying name given to an element of a program, such as a data type, a data structure, a procedure, a function or an individual statement. Depending on the programming language, an identifier may have to be 'declared' before it can be used within the program, at which stage its characteristics are specified (see *data declaration*). Sometimes referred to as symbolic name or label.

idle time The time during which the processor of a computer system is in operation but has no useful work to undertake.
Compare *down time*.

IEEE Abbreviation of *Institute of Electrical and Electronic Engineers*.

IF statement A program statement used in some high-level languages for a single or double branch. For a single branch it takes the form 'IF expression THEN statement', and for a double branch 'IF expression THEN statement1 ELSE statement2'. The statement in the THEN clause is executed if the expression is true and otherwise, if it is included, the statement in the ELSE clause is executed. In either case, 'statement' may consist of a compound statement.

IFIP Abbreviation of *International Federation of Information Processing*, an umbrella organisation for bodies representing computer specialists, based in the US. It runs regular conferences and publishes transcripts of the proceedings that are an authorative reference source on computer science and practice.

IGES Abbreviation of *initial graphics exchange specification*.

IKBS Abbreviation of *intelligent knowledge-based system*.

illegal character A character whose value is not permitted either in terms of the character set in use or the circumstances in which the character is being used. A transmission control character sent to a printer, for example, would be regarded as illegal, whereas it could legally be sent over a transmission line.

illegal operation An operation specified in an instruction that a processor regards as invalid and cannot execute. This may be because the operation code does not exist, because the operands have invalid values, or because the address is outside the memory limits set for the program.

ILS Abbreviation of *interactive learning system.*

image Information (usually digital) representing the visual appearance of an object. Thus a screen image represents the appearance of a display screen, a document image the appearance of a document. Contrast with *data* and *text*, where codes are used to represent the content of an object, rather than its appearance.

image grabber See *video scanner.*

image processing The use of computers to process images. It is made up of a number of disciplines:

(1) image enhancement – improving an image such as by increasing contrast;

(2) image restoration – reducing the effects of known degradation, such as noise or blurring;

(3) image analysis – calculation of statistical measures relating to certain objects within an image, such as the average particle size;

(4) image interpretation – extracting a small amount of descriptive information from the large amount of data contained in an image.

Image processing demands very large amounts of processor power and memory. It is used for machine vision; industrial inspection; digital (including medical) imaging; and for document storage and archiving.

See also *document image processing; image scanning.*

image scanning A process performed on a printed image to convert it into a digital form. The image to be scanned – a document or a technical drawing – is divided into picture elements (pixels) normally at a resolution of 200 pixels per inch (higher resolutions are used for engineering applications) and the blackness or whiteness of each pixel is measured and recorded as a stream of bits representing black (or sometimes grey scale values) and white dots within the image. The bit stream can be transmitted or stored electronically; can be used subsequently to reassemble a near-exact copy of the scanned image; or can be processed in order to recognise patterns within it – see for example *optical character recognition.*

immediate data See *immediate operand.*

immediate operand An operand carried within the instruction itself, rather than retrieved from a specified address elsewhere in memory.

impact printer A printer that produces characters on paper by physical contact, normally by striking an inked ribbon against the paper. Dot matrix, daisy wheel and line printers are examples of impact printers. Contrast with a *non-impact printer* such as a laser printer, which forms the image by electrostatic means.

import Transfer (usually data or documents) into one application or computer system from another. By implication, the data or document to be imported is not already in the correct format for the receiver, so will need to be converted into the right internal format on receipt. For example, many word processing packages on personal computers can import documents prepared using other popular word processing packages.

Compare *export*.

impurities See *dopant*.

in-house development Where a computer application is developed by an organisation's own information systems department, rather than being developed, for example, by a software house, or purchased in the form of an applications package.

in-house software See *in-house development*.

in-line coding Statements that are included in the main logic path of a program, rather than being called on from there when needed, such as in the form of a subroutine.

inclusive OR See *OR operation*.

incremental dump A file dump in which only the changes since the previous dump are recorded, rather than the full contents of the data files. Normally a full dump is taken periodically, say every weekend. Incremental dumps are taken in between, say daily, with the advantage that they take less time to carry out than a full dump. To recover the data files to their state on a given day, first the full dump is reloaded, then changes recorded on the subsequent incremental dumps are applied in succession.

index (1) For data files on direct access storage, a table of references to data records, also held on direct access storage but separately from the data records. Normally, each entry in an index consists of the key value of a record plus the address of the block in which it is held, and entries are held in key sequence. The file management software can search the index for a particular key value much more quickly than it can search amongst the data records themselves. To access the index still faster, it may hold some or all of the index entries in memory while the file is being processed. See also *indexed sequential file*.

(2) In programming, a number or address used for indirect addressing.

index page A page in a tree-structured database, such as in a videotex system, that directs the enquirer to other pages.

index register A register within a computer processor that is used by programs to address memory indirectly. Program instructions can load a value into the register. This value is added to the memory address used by certain instructions to derive the actual address referenced.

indexed sequential access method (ISAM) An operating software component that enables applications programs to create and access indexed sequential files. Also used, usually abbreviated to ISAM, to describe files so organised.

indexed sequential file A way of organising data files on direct access storage such as disk. Records are held in sequence according to one or more specified key fields within the record. Additionally, an index is created by the software, consisting of the key values of records in the file, together with the location of the record, or more often of the block in which it may be found. To find a particular record, the software searches the index, then reads the indicated block and searches it to find the record. Usually, blocks are only partially filled with records when the file is first created, to leave space for later insertions. If there is no space left to insert new records in sequence, they may be allowed to 'overflow' into overflow blocks. This gradually reduces processing efficiency and eventually the file may need to be reorganised.

Indexed sequential files have the advantage that records can be read in key sequence for batch processing and can also be retrieved directly for transaction processing or enquiry applications.

Compare *random file*.

indicator A data field used to record a condition that has arisen within a program. At a convenient point, the program tests the indicator and takes appropriate action. Also known as a *flag*.

indirect addressing A technique used by programs to address items in memory, in which the address contained in the instruction is the address of a memory location that, in turn, contains the address to be referenced. The technique is used in multithreading systems, where the same section of program may be used concurrently to process different data. This can be achieved by altering the address in the memory location referenced by the instruction, leaving the program instruction itself unchanged. It is also used in systems with dynamic memory management, where programs and data may be moved about in memory while they are executing – see *memory management*.

industry-standard architecture A standard for connections to a personal computer bus via expansion slots. This standard was derived from the specification for the AT version of IBM's PC, whose bus has a width of 16 bits and a speed of 8 MHz.

See also *extended industry-standard architecture*.

inertia See *systems inertia*.

146

inference engine The software component that drives a knowledge-based system such as an expert system. The inference engine simulates the deductive thought processes of a human expert. Based on the information provided by the user of the expert system, it interprets the rules and facts contained in the knowledge base to draw logical inferences about the problem that is being analysed.

informatics See *information technology*.

information (1) In its most precise sense, information occupies a place in a hierarchy between raw facts (which may be referred to as *data*) and knowledge. Data is transformed into information when it is put into context and related to a particular problem or decision. On this basis it can be defined as 'facts to which a meaning has been attached'. This meaning can only be attached by a human being, by a conscious act of cognition. To put this another way, information cannot be sent to someone, it can only be received, because until the recipient has recognised it as such it does not yet constitute information.

(2) Some information scientists argue that information may be considered a physical entity in its own right, just like matter and energy, although it clearly is not as 'real' as they are. They point out that all organised structures contain and may transmit information – DNA, for example, from the organic world, or a silicon chip from the inorganic world.

(3) The term is widely used in a more general sense than either of the above definitions, as in *information technology* or *information processing*, to encompass all the different ways of representing facts, events and concepts within computer-based systems. Thus it embraces forms such as data (i.e. numbers and structured text), text (as in documents), image and video. (It is used in this sense in other definitions in this dictionary, not because this is a 'preferred' meaning but rather because no other suitable term exists.)

information centre (IC) A corporate support centre for end-users. Originally, information centres were set up to help managers in particular to obtain the information they required from data files held on centralised computers. Information centre staff would either use powerful software tools such as enquiry languages and report generators to obtain the required information, or would help end-users to do this for themselves. With the arrival of the personal computer, the role of most information centres broadened to provide support for personal computer hardware and software used by a range of office staff, and also to educate and train staff to get the most from the equipment.

information executive See *chief information officer*.

information hiding A concept associated with object orientation. It specifies that the private data and functions of an object cannot be accessed from outside directly, but only in a controlled manner via its publicly known features.

See also *encapsulation*.

147

information management A general term used to characterise any form of business or office management that considers information (data, text, image, etc.) and information resources (including personnel and information systems) as primary assets. The term is often used with a narrower meaning, referring to methods used to manage information, and particularly data, within a computer-based system – see *data management*.

information overload Where an organism receives more signals in the form of information than it can assimilate. The term is sometimes used to describe the situation that arises when computer-based systems are introduced into an organisation and generate more information than the people using them can cope with. Often, the problem is not so much that too much information is provided, but rather that the information generated is not sufficiently refined, so that its recipients have to search through large amounts of irrelevant information to find what they want, or have to spend a great deal of time summarising and recasting information to get it into a useful form. This situation has been neatly summarised in the phrase 'too much data, too little information'.

information policy Rules applied within an organisation, and particularly within large multi-division or multi-company organisations, intended to regulate the way computer-based systems are introduced and used. Information policy might define, for example, how business units are to be charged for the use of centrally-provided systems and services; which systems are to be common throughout the organisation; what standards are to be observed when procuring equipment or developing applications.

information processing A term that superseded *data processing* when the application of computers widened beyond the basic operational systems of organisations that were the initial targets. Thus it includes the application of computers to tasks in the office, the warehouse, the sales outlet, the factory; and the processing of documents, images, etc. as well as structured data.

information processing department See *information systems department*.

information provider An organisation that owns and originates the information accessed via an information retrieval service such as videotex. Contrast with *service provider*, who acts as a kind of wholesaler, providing the means to store and retrieve information obtained from a number of different information providers.

information rate See *data transfer rate*.

information resource dictionary system (IRDS) A synonym for data dictionary or encyclopaedia. The American National Standards Institute (ANSI) uses this as the name for a standard for data dictionaries which it is developing.

information resource management (IRM) A view of information as a resource to be managed like any other business resource (such as personnel or finance). The US Paperwork Reduction Act defines it as 'the planning, budgeting, organising, directing, training, promoting, controlling, and other managerial activities involved with the creation, collection, use, and dissemination of information'.

information retrieval The use of an online terminal to search files of information, held by a computer system, in order to locate, then display or print particular items. Information retrieval applications can be divided into two broad categories:

(1) Data retrieval, where the files consist of codes, numbers and structured text, such as details of customers, orders, suppliers, personnel, accounts and so on, often organised according to database principles. In this case, queries specifying search criteria may either be keyed in at the terminal using a query language or may be entered by selecting predefined questions built into an applications program.

(2) Text retrieval, where the files consist of free form text, such as documents, reports or abstracts. Here, searches may be based on specified keywords or on combinations of words in the text (the latter is known as *full text retrieval*).

Recently, image retrieval systems have started to be introduced, which allow images – of documents, drawings, maps – to be retrieved using any of the search techniques already mentioned.

information security administrator See *data security officer*.

information society A society in which the key products consist of information, and hence the key resources are ideas and knowledge. It is also called the post-industrial society, and can be contrasted with an industrial society, in which the key resources are capital and the key products are manufactured goods. By extension, in an information society business organisations will succeed and add value to their products and services through their ability to exploit ideas and knowledge, rather than through their ability to exploit manufacturing capacity.

information system (IS) In general, any system that processes information, for example an educational system, an intelligence-gathering system. In the context of information technology, it means a system within an organisation that processes and distributes the information the organisation needs to plan, monitor and control its activities. These information systems are operated by, and therefore include, people as well as technology. They use a number of well-established technologies, such as paper filing systems and typewriters, although information technology is playing a growing and often a dominant role.

The term is commonly used in a narrower sense to refer to information systems based on information technology, in other words as synonymous with *computer-based system*. See, for example, *information systems department*. Compare with *computer system*, which is an essential component of a computer-based system, but only one part of it.

information systems department The department within an organisation responsible for designing, building, operating and maintaining computer-based systems, or some subset of those activities. Sometimes abbreviated to IS or MIS department. Has begun to supersede the less fashionable earlier designations 'data processing department' or 'computer department', reflecting the fact that the scope of computer-based systems has broadened well beyond data processing and computers alone.

information technology (IT) Electronic technologies for collecting, storing, processing and communicating information. They can be separated into two main categories: (1) those which process information, such as computer systems, and (2) those which disseminate information, such as telecommunication systems. Increasingly, the term is used to describe systems that combine both.

As an industry and in terms of applications of the technology, information technology borders on and overlaps a number of longer established technologies, such as office equipment, manufacturing, broadcasting and publishing. Information technology is an accepted term in the UK but is not a universal term. Terms like telematics (widely used in France) and informatics are also used elsewhere.

information theory The science of messages. Information theory is based on mathematics, and deals with the factors which affect the transmission of messages, such as channel capacity, noise and information content. The discipline was established by Claude Shannon of Bell Laboratories, whose ground-breaking paper 'A Mathematical Theory of Communication' was published in 1948.

inheritance A concept whereby generic attributes of a particular class of objects (for example, entities in a database) are defined only once, and are implicitly assumed to be inherited by (in other words apply to) all members of the class. The concept is particularly associated with object-oriented databases and object-oriented programming, in which context it is also referred to as specialisation. From a programming viewpoint, inheritance means that a new object can easily be derived from an existing object, by defining it as a subset or superset of its 'parent'.

See also *class hierarchy*; *generalization hierarchy*; *polymorphism*.

initial graphics exchange specification (IGES) A draft standard for the format and content of information describing products in engineering and graphical terms, so that this can be exchanged between different computer systems (formulated by ANSI).

See also *RIB*.

Compare *graphics kernel system*; *PHIGS*.

initialise Set to an initial state, ready for execution or use.

(1) Used of a program or of a routine within a program, it means to set all the indicators and variables to the required initial values.

(2) Used of a disk, it means to run a special initialisation program. Before they can

be used to store information, disks normally have to be initialised by the computer system on which they are to be used. During the initialisation process, markers are written on to the disk to show where sectors start; alternate tracks are allocated to replace any that are found to be faulty; space is reserved for a directory of free and allocated space on the disk; and the disk is given an identifier, such as a name or serial number.

See also *group-coded recording*; *modified frequency modulation*.

ink jet printer A printer that creates the print image by squirting ink drops on to the paper selectively from a number of tiny nozzles. The ink drops are passed through an electrical field which directs them precisely on to the paper. Ink jet printers are quiet and can produce high-quality print (resolutions up to 360 dots per inch), both in black and in colour. They are less expensive than laser printers, but also slower and fussier about the quality of paper used.

input As a verb, (1) transfer data or programs into the memory of a computer or (2) enter data into a personal computer or a data terminal, via a keyboard or by some other means.

Also used as a noun to refer to the information that is transferred in either of the two ways described above.

input area See *input buffer*.

input buffer An area of memory reserved for incoming blocks of records from a data file. See *buffer*.

input devices Devices used to enter data (and sometimes programs) into a computer system. In the past, card readers and paper tape readers were widely used for this purpose, but they have been displaced by keyboard devices used by people to enter data directly, such as visual display terminals. Increasingly, scanner devices are also being used that read and input data already prepared in some form, such as on paper. For graphics applications, a light pen, a graphics tablet or a mouse may be used.

input record A data record from an input file, such as a magnetic tape file, transferred into a program's data area ready for processing.

input/output (i/o) A general term meaning the transfer of information between the memory of a computer and its peripherals and terminals.

input/output channel (i/o channel) A component of larger computers (small computers usually have a bus, rather than channels) that carries data between peripherals and memory. These channels come in two main types – selector channels that carry bursts of data from high-speed peripherals such as disk drives; and multiplexor channels that carry the data from a number of slower peripherals such as printers or terminals.

input/output interrupt (i/o interrupt) A signal to the processor from a peripheral, or from a peripheral controller or interface, that a transfer has completed, or that some other event has taken place that requires the processor's attention.

See also *interrupt handling*.

input/output limited Where the throughput capacity of a computer system is determined by the input/output hardware, and usually by access to data files on disk.

Compare *processor limited*.

input/output logic A subsystem that links the peripherals that bring information into and out of a computer system with the processor and memory that process the information. Its role is to match up the varying speeds of the peripherals with that of the processor and memory. As the controlling unit, the processor must of necessity operate faster than any one of its peripherals, and must also be capable of controlling a number of them in parallel. Technologies used in the input/output logic include channels and the bus.

See also *interrupt handling*.

input/output subsystem See *input/output logic*.

install Do whatever is necessary to make available for use. Used both of complete computer systems, and of their components or features. Installation of a large-scale computer system involves preparing the computer room; physical installation of the equipment; interconnection and testing of the various component; then loading and testing of operating software. Installation of a personal computer is a much simpler business, and installation of a new software package or feature may simply be a matter of copying a disk and/or running an installation program.

installation A place where one or more computer systems have been installed. It is normally used of shared rather than personal computers, in other words of places where computer specialists run computers for the benefit of other people.

installation standards The procedures and ways of working adopted at a particular installation. Installation standards are normally written down in a formal document, specifying, for example, how computer errors and failures are to be recorded; how and when the suppliers' maintenance engineers should be called; how programs are to be documented.

instantiate Provide a specific example of a generalised example or concept. The term is used in logic programming, in which variables are used to stand for objects that are yet to be determined. When the object for which the variable stands is known, it is said to be instantiated.

Institute of Electrical and Electronic Engineers (IEEE) The professional association for US engineers. Working with the US national standards authority, ANSI, it has set up a number of special groups to define standards for telecommunications. Group 802, for example, originated standards for local area networks that have since been adopted by the international standards body, ISO.

institutional systems See *operational systems.*

instruction A command to the processor of a computer system. At the lowest level, an instruction is the smallest unit of a program and is executed directly by the processor. Normally, instructions are originated by programmers in the form of statements in a programming language that are subsequently translated by a compiler or interpreter into the instructions executed by the processor. Generally, instructions have up to three components – a) an operation code, b) an operand, and c) an address. They take the form do 'a' to 'c' with 'b', for example 'store this register into memory location x', or 'transfer control (branch) to the instruction at memory location y'.

instruction cycle The operations by the control unit of a processor to perform a program instruction. The instruction cycle consists of a fetch cycle and an execute cycle.

instruction decoder The part of a processor that decodes program instructions before they are executed. See *control unit.*

instruction execution See *execute cycle.*

instruction execution time See *execution time.*

instruction fetch See *fetch cycle.*

instruction format The way in which the various components of a program instruction are laid out. Program instructions may have up to three components – an operator or instruction code, one or two operands, and an address. The instruction code always occupies a fixed position in the word or words that each instruction occupies. The position of the other components within the instruction, and sometimes the overall length of the instruction, are fixed for each type of instruction, but may vary from one type of instruction to another.

instruction path length The number of instructions that are executed when a particular route through a program is followed.

instruction set The set of instruction codes (or operators), each represented by a different bit pattern or code, that a particular computer accepts as valid and is capable of executing. The instruction set of a modern computer includes instructions to carry out a range of arithmetic calculations; instructions to move data about in memory, between memory and registers used for calculations, and within the registers; instructions to make comparisons and tests; instructions to branch to other points in the program; and instructions to initiate transfers of data between memory and peripherals.

integrated Forming a complete and coherent whole. Unfortunately this word has been overused by many people in the information technology industry and sometimes has no real meaning at all. It is used in a special sense in integrated package and integrated circuit.

integrated application architecture See *application architecture.*

integrated circuit (IC) A device made from semiconductor materials (usually silicon, based on sand) that is manufactured containing many interconnected circuit elements on a single 'chip' (a piece of silicon a few mm. square). Integrated circuits, popularly known as silicon chips, are used both for computer processors and for memory, as well as in a wide range of domestic and industrial products.

Semiconductor materials were first used to produce transistors, which replaced the thermionic valves used in the first generation of computers. At first, transistors were produced as individual electrical components. Hundreds of them were fabricated on a single slice (known as a wafer) of silicon, but they were then physically separated and assembled individually with wires and a protective housing, so that they could be mounted along with other elements such as resistors and capacitors in electrical circuits.

Progress took a giant step forward in 1958 when the integrated circuit was invented separately by Jack Kilby of Texas Instruments and Robert Noyce, then of Fairchild Semiconductor. On an integrated circuit, a number of transistors and other circuit elements are interconnected electrically. Impurities or dopants are introduced into the semiconductor material to alter its electrical characteristics. This creates negative (n) and positive (p) regions, which are sandwiched in p-n-p or n-p-n form to produce transistors. Integral resistors are constructed by using the body resistance of the semiconductor itself, while the inherent capacitance of the junctions between the p and n regions serves to provide the capacitors. This means that an entire electronic circuit, such as a processor or a section of memory, can be produced at one go. Hundreds of such circuits are fabricated on a single wafer of silicon, then they are separated and tested. Faulty circuits are discarded, and the remainder are provided with wire 'feet' and a covering of plastic material so that they can be mounted on printed circuit boards. The first integrated circuit held several transistors, but the number soon grew to hundreds and now run into millions.

See also *wafer scale integration.*

integrated office systems See *office automation.*

integrated package An applications package, usually for a personal computer, which includes a number of related applications. Integrated packages are so designed that it is easier to exchange information and switch between the applications they include than if separate applications packages were used for each. The first integrated package, and also the most successful so far, is Lotus Corporation's 1-2-3, which combines spreadsheet, data management and business graphics.

integrated program support environment (IPSE) A set of computer-based tools for specifying, designing, programming and testing computer applications. IPSEs are intended to provide a complete method for developing systems that is independent of the programming language used.

integrated services digital network (ISDN) A term coined by telecommunications authorities to describe digital public networks that provide both voice and data services over a single connection, often abbreviated to ISDN. It is used both to describe the networks themselves and the standards that define how they will operate. A single connection to an ISDN network, known as basic access, such as might run to a private house, consists of two digital channels each running at a speed of 64 kilobits per second, plus a 16 kilobit/second channel to carry control signals. Business users will be offered primary access connections, consisting of 24 64 kilobit per second channels, one of which is used for control signals. The channels can be used either for digital voice or for data.

integrity Used of a set of data, such as one or more data files, to mean that it is internally consistent and represents what it is supposed to represent.
See also *referential integrity*.

intellectual property Ownership rights relating to the intellectual content of products such as software packages. Intellectual property rights, reflected in copyright laws, are intended to protect authors and publishers of software against the risk of being denied due reward for their efforts, firstly because software distributed on floppy disk is so easy to copy physically, and secondly where their ideas are stolen for incorporation into competitive products.
See also *copy protection*.

intelligent Used of computer-based equipment in general to mean capable of running a program, and thus able to take its own decisions as to how information should be handled. Hence, for example, intelligent terminal. Used in a different sense by the artificial intelligence community, meaning capable of simulating some of the processes of human thought.

intelligent building An office building designed to provide its occupiers with advanced services based on information technology. Most buildings claimed by their developers to be 'intelligent' include computer-based energy management and security systems; are already wired up for telephone and data communications; and offer shared tenant services such as word processing centres and videoconferencing rooms.

155

intelligent character recognition (ICR) A technique for recognising characters electronically that is more sophisticated than conventional optical character recognition. A character is identified by examining its shape and the features of the strokes that are needed to create it.

See also *feature extraction*.

intelligent front end An addition to an existing computer application, designed to make it easier for people to use. A natural language interface, for example, serves as an intelligent front end to a conventional database, enabling users to express queries in an English-like language rather than an obscure formal syntax. As does a natural language interface, an intelligent front end normally uses inference procedures characteristic of knowledge-based systems, hence it is also known as *knowledge-based front end*.

intelligent knowledge-based system (IKBS) Synonymous with *knowledge-based system*.

intelligent terminal A terminal which can store data and process it before it is displayed or printed, responding to commands included in the data or using a program stored in the terminal itself. Contrast with *dumb terminal*, which cannot store or manipulate the data it receives, but can only send it directly to a display or printer. Also known, particularly in the US, as *smart terminal*.

intelligent typewriter A typewriter that can store and retype text held in an internal memory or taken from a removeable storage medium such as magnetic card. These devices are particularly appropriate for repetitive texts and revision typing. Except for these applications, however, they do not give the kind of productivity gain that can be achieved with word processors with a display screen.

interactive Where the user of a system is able to conduct a dialogue to solve a problem, rather than entering or sending off messages and only receiving results or a reply later on. Thus the telephone is an interactive communication system whereas telex and the postal system are not, and BASIC is an interactive programming language whereas COBOL is not.

interactive debugger A software tool that can be used to trace and correct errors in a program while the program is being run. It is loaded into memory alongside the program under test, and allows the programmer to control the operation of the program and inspect the contents of memory.

interactive learning system (ILS) A system that encourages the exchange of knowledge or information between people involved in some form of dialogue. The term is used both for systems involving people alone and for systems based on technology, and particularly on computers. Interactive learning systems based on computer technology may be used by a single user individually (see *interactive video* and *interactive multimedia*); by a group of users within a classroom; or by geographically-distributed users communicating over a network (see *computer conferencing*).

interactive mode A method of operating a data terminal where a user is able to communicate directly with the application he or she is using in a computer system, so that it replies immediately to the messages entered via the terminal. Also known as *conversational mode*.

interactive multimedia A term used to describe applications, usually based on personal computers, that present the user with a combination of moving pictures and/or sound, along with text and graphics displayed on the computer screen. The user is able to control the sequence in which this material is presented by interacting with the controlling computer.

See also *hypermedia*.

interactive video The combination of a computer system and a videodisc player, so that still and moving video images can be displayed under the control of the user via the computer keyboard. At its simplest, the computer can serve as an index to a picture library, searching a database of details about the pictures on the disk for matches to queries. When a match is found, the computer sends the frame number of the required picture to the videodisc player, which finds it and displays it on the screen. Applications range from textile collections to blood samples and libraries of works of art. It is also possible to combine computer-generated images with the video images – see *genlock*.

interactive voice response (IVR) Applications where a computer responds automatically to telephone calls. Callers instruct the computer using a limited vocabulary of commands or by tones generated by a touchtone telephone, and it responds using pre-recorded or synthetically produced speech.

Compare *audiotex*.

interblock gap The empty space between successive blocks on a tape. This space is there to allow the tape to slow and accelerate between blocks, so that the tape is always moving at the right speed when the computer wishes to read or write a block.

interblock space See *interblock gap*.

interconnection See *open systems interconnection*.

interface The meeting of two components of a computer system or of an information system that have differing characteristics. Thus they must be matched in order for the system as a whole to operate. The term is used both for the manner in which the matching takes place (as in 'man/machine interface', now often replaced by the non-sexist 'human/computer interaction') and for the device that executes it (as in 'printer interface' or 'terminal interface'). Two components that need to interface may differ in terms of electrical characteristics such as the speed at which they transfer data. This applies, for example, to the connections between a processor and its peripherals, all of which transfer data at a slower rate than the speed of the processor. They may also differ in terms of the methods used to communicate or represent information, which is obviously the case when people use computers.

See also *standard interface*.

interlace A mechanism to reduce the effective time needed to retrieve data from computer memory. Memory is divided into a number of sections, each held on a separate chip. Rather than each chip containing a continuous section of memory in addressing terms (for example words 0 to 1,023), each chip contains one or two bits from each word for a much larger section of memory. When it requires data from memory, the processor, whose cycle time is faster than that of the memory, accesses one chip in its first cycle, another in its second, a third in its third, and so on. It still takes, say, four processor cycles to obtain a word of data from any one memory chip, but it will take only seven processor cycles, rather than sixteen, to obtain four words of data from four different chips.

interleave Access (normally hardware) resources in other than normal physical sequence. Disk drives, for example, normally interleave sectors on the storage media so that they can read successive sectors in one pass of the read/write heads. Without this, i.e. if the sectors were arranged in sequence physically, the drive might not have enough time to read successive sectors and consequently access time would be longer.

intermediate storage See *work area*.

internal memory See *memory*.

internal schema Used in database parlance to mean a schema (a programmed description of entities and of the relationships between them) representing data as it is recorded physically on the storage medium. Also known as *physical schema* or *storage schema*.

Compare *conceptual schema*; *external schema*.

internal storage See *memory*.

International Data Exchange Association See *electronic data interchange*.

International Standards Organisation (ISO) An international standards-making authority, operating as a 'non-treaty' body of the United Nations. Policy is decided by representatives of national standards bodies such as Britain's BSI and ANSI from the US. It defines standards for a wide range of products and services, including in information technology. The latter are under the control of a technical committee formed jointly with the International Electrotechnical Commission. Much of its current work within the field of information technology centres round the reference model for *open systems interconnection*, often referred to as the OSI model.

International Telecommunications Union (ITU) A special agency of the United Nations whose purpose is to promote international cooperation in telecommunications, and to harmonise national interests. CCITT operates under it as an international standards-making body, responsible among others for the V-series and X-series recommendations covering data communications.

Internet (1) A global research network, consisting of a loose confederation of interconnected networks. In 1992 it linked over 950,000 computers attached to 8,000 networks in 80 countries. It provides services such as file transfer and electronic mail.

(2) Also used (without initial capital) to mean the equipment interconnecting different networks, as opposed to the equipment forming them. See also *TCP/IP*.

interoperability The ability of a software or hardware component to operate in conjunction with other components, independently of their origin. For example, a fourth generation language that could be used with a wide range of database management systems from different suppliers could legitimately be described as having high interoperability.

Compare *portable*.

interpreter A program which takes as input a program in a high-level language (such as BASIC) and translates and executes the statements in one operation. Contrast with *compiler*, which translates source program statements into an object program in one operation, and which must then be executed as a separate operation. The advantage of an interpreter is that programs can be developed in a highly interactive way, but programs executed in this way make much less efficient use of computer resources than those that are compiled.

interrecord gap See *interblock gap*.

interrupt A signal to the processor of a computer system generated by one of its peripherals or by some other device under its control, such as a clock circuit. An interrupt warns the processor that an event has taken place that requires attention. A peripheral generates an interrupt, for example, when a data transfer to or from it terminates, so that the processor can check that all is well and, if required, initiate a further transfer.

interrupt address vector See *interrupt vector*.

interrupt handling Logic within a computer processor and its operating system that takes appropriate action following an interrupt from one of its peripherals or from other devices such as the real-time clock. A peripheral generates an interrupt, for example, when a transfer of data is complete. When it receives an interrupt, the processor temporarily suspends the sequence of program instructions it is executing at the time, gets the status information that defines what device generated the interrupt and why, then passes control to an interrupt service routine. It does this by reference to an interrupt vector – a list of the addresses of interrupt handling routines to be initiated depending on the identity of the interrupting device. These interrupt service routines signal to the interrupting device that the interrupt has been serviced, and store any status information for future reference, such as by the program that originally initiated the transfer.

In more complex systems capable of running a number of programs concurrently (multiprogramming or multitasking systems), they then update the status of any program affected by the interrupt. The program that initiated a transfer, for example, may be suspended waiting for it to complete and can now be reactivated. The processor then passes control to the highest priority program that is active. Thus if a higher priority program becomes active as a result of an interrupt, the program suspended when it arrived will have to wait its turn until it reaches the top of the priority list again. This makes sure that different programs competing for the use of the processor get a fair share of its attention.

interrupt priority level The priority accorded to an interrupt received by a processor from an external unit such as a peripheral. See *interrupt handling*.

interrupt service routine A short program that is activated to carry out the processing needed immediately an interrupt is received by a processor from an external unit. This will include storing the current status of registers, so that the program suspended when the interrupt was received can resume at the point at which it was interrupted.

interrupt signal See *interrupt*.

interrupt vector A list of addresses used by the processor to identify the interrupt service routine to be activated when an interrupt is received from an external unit such as a peripheral. See *interrupt handling*.

interworking Where two or more computers exchange information directly, to work on a task in cooperation.

inverse video See *reverse video*.

invert Of a graphics image or a section of a display screen, change it so that white is seen as black and vice versa.

inverted file A form of file structure in which every item (or every major item) in each record is represented in the index. This makes it possible to retrieve any record in the file meeting given criteria relatively quickly. Inverted file structures are used for files which are searched using a variety of criteria, such as files of text abstracts which may be searched for records containing any combination of keywords. Inverted files are costly to create and store, which offsets the advantage of rapid access.

See also *associative retrieval*; *full text retrieval*.

inverter A circuit with one input whose output is high if the input is low and vice versa.
See also *logic element*.

invisibles Characters that cannot be printed and that word processing software normally does not display on the screen (some packages permit this as an option to make editing easier). This includes, for example, RETURN and TAB characters.

IPSE Abbreviation of *integrated program support environment*.

IRDS Abbreviation of *information resource dictionary system*.

IRM Abbreviation of *information resource management*.

IS Abbreviation of *information system*.

IS policy See *information policy*.

ISAM Abbreviation of *indexed sequential access method*.

ISDN Abbreviation of *integrated services digital network*.

ISO Abbreviation of *International Standards Organisation*.

isochronous transmission See *synchronous transmission*.

issue-based information system (IBIS) A method of representing and addressing complex problems on a computer. The method was developed by Horst Rittel at the University of Stuttgart, and views complex problems (termed 'wicked problems', where the problem itself is poorly formulated) as a conversation between stakeholders. The computer system helps them to express the problem and mediates their team effort to solve it.

IT Abbreviation of *information technology*.

IT function See *information systems department*.

iterative Used of a routine that performs a series of operations repeatedly until a desired

161

result is obtained, or of a process that is repeated until results are acceptable.
Compare *recursive*.

ITU Abbreviation of *International Telecommunications Union*.

IVR Abbreviation of *interactive voice response*.

J

Jackson method A structured programming method originated by a UK programming specialist called Michael Jackson.

JANET Abbreviation of Joint Academic Network, a packet-switching network used by UK academic institutions.

JCL Abbreviation of *job control language*.

JIT Abbreviation of *just in time*.

job A unit of work for a multiprogramming computer system, consisting of a program run or a series of related runs. Normally, each job has a single set of operating instructions, which may be expressed in a job control language.

job control language (JCL) A language used by programmers to instruct the operating system of multiprogramming computer systems how programs are to be run – what resources and storage media they will require, what action to take if certain conditions occur, etc.

job scheduling The process used by the operating system of a computer system capable of multiprogramming (running a number of programs concurrently) to make sure that its resources are used as efficiently as possible and that programs receive an appropriate share of those resources, according to their relative priorities and the nature of the processing they carry out. Thus a program responding in real time to messages entered at terminals will have priority over a program running in batch mode; while if two different programs wish to use a single serial resource, such as a high-speed printer, the higher priority program will have first use. Job scheduling may be driven by any or all of the following:
   (1) the general requirements of the user organisation, specified when an operating

163

system is installed;

(2) the characteristics of the programs that are running, entered in the form of job control language statements when the program is started;

(3) commands entered by the computer operators.

join An operation (based on set theory) performed on a relational database that links two tables to form a third.

See also *relational model*.

Josephson junction A semiconductor device combining fast switching with low power dissipation. It is based on a junction between two metals which becomes superconductive at temperatures approaching absolute zero.

journal See *transaction journal*.

Joy's Law A theory advanced by William Joy, co-founder of Sun Microsystems, predicting the speeds that will be attained by processor chips. In 1984, he stated that processor speeds would increase exponentially according to the formula:

$$Mips \ (millions \ of \ instructions/second) = 2*(year-1984)$$

This implies that speeds will exceed 65,000 Mips by the year 2000.

joystick A small device consisting of a lever terminating with a ball which can be moved through an angle of approaching 90 degrees from the vertical in any direction. Used to control the cursor of a visual display, mainly for games applications on home computers.

JPEG algorithm The JPEG (Joint Photographic Experts Group) algorithm is used to compress data such as colour images where some loss of information, and hence loss of quality in the compressed image, is acceptable. Compression ratios of between 2:1 and 20:1 can be achieved, depending on the acceptable loss of quality.

Compare *MPEG algorithm*.

jukebox A device that holds optical disks until they are required for reading. Typically, a jukebox can hold several hundred optical disks. It operates just like a jukebox for gramophone records in that it can be instructed by a computer system (or by some other device to which is connected) to select any one of the disks it contains and bring it to its reading station, where information can be read from it and transferred across to the requesting system.

jump instruction See *branch*.

just in time (JIT) Precision scheduling of supplies of manufacturing components and parts, designed to minimise the holding of stocks.

justify (1) Adjust a data field in memory to align with one end of a larger field – right justify to put it at the right hand end and left justify to put it at the left hand end.

(2) Adjust the space between letters in a line of text so that they run to the end of a defined width. As a result, the margin appears straight, as opposed to ragged right or ragged left margins.

K

_____

K Abbreviation of *kilo-*.

Kb Abbreviation of *kilobyte*.

kbit See *kilobit*.

kermit A versatile file transfer protocol designed for use over public telephone networks and supported by many bulletin boards and computers of all sizes. It can use a number of methods of error detection and flow control. Two different versions of Kermit running on different systems can 'negotiate' a mutually suitable method of operation.

kernel The 'heart' of an operating system. It contains those functions that are specific to the machine on which it is running, such as device drivers and routines handling interrupts and basic input/output tasks.

kerning Adjusting the spacing between a pair of characters to improve their appearance and legibility. A term borrowed from the publishing industry and used in word processing and desktop publishing applications. Some word processing packages do it automatically based on tables of pairs of characters held in the computer, and others allow the user to adjust the spacing.
Compare *tracking*.

key (1) A designated field used to identify a record in a data file, and particularly to identify where it is located within the file. For example, the customer code might be the key in a file of records containing customer details, or surname and initials in a file of mailing addresses. The key value may either be held in the index to the file (indexed sequential files), or may be used to calculate directly the address of the block where the record is stored (random files).
(2) A lever or button on a keyboard, which can be touched or depressed to generate a

character.

key punch A machine with a keyboard used to record data in punched cards or paper tape.

key-to-disk unit A device used for data preparation, consisting of a keyboard, a small display, and a disk drive. Data entered at the keyboard can be verified, then stored on disk or tape for processing later.

keyboard A device used to enter text into a computer system or terminal. It works by generating character codes according to which key, or combination of keys, is pressed, and sending those codes to the processor of the device to which it is attached. Most keyboards are like typewriter keyboards, with the addition of a number of special keys, such as cursor control and function keys. Keyboards with a more limited range of characters are also used for special purposes, such as numeric keypads used for accounting applications. On modern systems, keyboards are usually separate units, attached to the processor or display by a flexible cable.

keyboard send/receive (KSR) A terminal with an alphanumeric keyboard and a printing device, used to send messages typed on the keyboard and to print out messages received from elsewhere. Terminals such as this are used as remote printers and as data entry terminals.

keyboard terminal See *keyboard send/receive*.

keypad See *numeric keypad*.

keyword retrieval Retrieval of text from a file (of documents or abstracts) using a specified range of keywords to identify the records that are required. Keyword retrieval can be contrasted with full text retrieval where any combination of words (or part words) contained in the text can be used as search criteria.

keyword search See *keyword retrieval*.

kilo- (K) A prefix meaning thousands of, as in kilobits or kilobytes. Often abbreviated to K.

kilobit A unit of data volume. In kilobits per second, a measure of the speed of transmission of digital data. For example, a digital telephone line transmits the digitally encoded speech signals at a speed of 64 kilobits per second (abbreviated to 64kbit/s or 64kbps).

kilobyte (Kb) A measure of storage volume, meaning 1,000 bytes. It is often abbreviated to Kb. When used to measure the capacity of computer memory, however, it normally means the first power of 2 exceeding 1000, which is 2^{10} or 1024. This is because memory chips are made of a round number of bits in binary, such as 2^{16} (65,536). Thus a personal computer described as having 64 kilobytes (or 64Kb) of memory in fact has 65,536 bytes.

KiloStream British Telecom's trade name for its point-to-point digital transmission services. These are intended for simple low speed data or for simultaneous voice and data transmission, operating at speeds from 2.4 kilobits per second (kbit/s) up to 64 kbit/s. Also available in multiples of 64 kbit/s for applications such as video conferencing or interconnecting local area networks.

kludge A makeshift correction or 'fix', usually to a program.

knowledge acquisition facility The part of an expert system that conducts a dialogue with a human expert (or a knowledge engineer acting as intermediary) in order to acquire and encode the knowledge base. Thus it provides a communication channel between the human expert or knowledge engineer and the expert system itself.

knowledge base An assembly of facts agreed by experts; the common knowledge they have acquired over years of work; and the rules of thumb (heuristics) that they apply to derive conclusions. This is so organised and encoded that it can be interrogated via an expert system.
See also *inference engine*; *knowledge acquisition facility*.
Compare *database*.

knowledge-based front end See *intelligent front end*.

knowledge-based system A system which applies a stored representation of human knowledge to perform a task, using inference procedures (see *inference engine*). An expert system is a type of knowledge-based system. Also known as IKBS or *intelligent knowledge-based system*.

knowledge elicitation A systematic process used by a knowledge engineer to discover the knowledge of a human expert, so that it can be incorporated in an expert system. A number of techniques are used, ranging from conventional structured interviews to more contrived techniques such as the repertory grid, where experts are presented with elements from the domain under study and are asked to define the relationships between them. Statistical analysis is then used to reveal new relationships that are implicit in what the experts have said.

knowledge engineer A person trained to acquire knowledge from a human expert so that it can be incorporated into an expert system in the form of a knowledge base. In other words, a knowledge engineer acts as an intermediary between a human expert and an expert system, encouraging the expert to articulate the rules and principles associated with his or her problem-solving skills. The knowledge engineer uses a specialised system of logical constructs to express the expert's knowledge in a useable form for the expert system.

Experts are often not conscious of the rules that they apply intuitively, and also find it difficult to express these in meaningful terms for a non-expert, so it may take a knowledge engineer many years of effort to extract and translate a particular body of expert knowledge.

knowledge processing In its established sense, knowledge is defined as familiarity or understanding gained through human experience. But the term is used in a different (and some would say suspect) sense by the artificial intelligence community, who use it to mean information so organised that it can be processed by computers in a way that mimics some of the processes of human thought, terming this *knowledge processing*. See, for example, *knowledge base* and *knowledge elicitation*.

knowledge representation Methods for expressing knowledge so that it can be stored and processed by a system with artificial intelligence such as an expert system.

knowledge representation language (KRL) A high-level programming language for expert systems. It enables an expert or a knowledge engineer to write the rules and describe the objects to be included in the knowledge base, simply and concisely.

KRL Abbreviation of *knowledge representation language*.

KSR Abbreviation of *keyboard send/receive*.

L

---

label An identifier associated with a statement in a program. Other statements in the program can use the label to reference the statement concerned, such as to branch to it.

LAN Abbreviation of *local area network*.

landscape format Used of pages printed by a word processing application where the width of text (or other information) printed on the page is greater than its length. Contrast with *portrait format* where the reverse applies.

LAP-B Abbreviation of *link access protocol*.

lap portable See *laptop computer*.

laptop computer A personal computer that is small enough to be carried around in a briefcase and to be operated on the lap. Unlike earlier suitcase-sized portable computers which relied on mains power and sometimes weighed over 10 kilos, laptops are battery-driven and weigh 6–7 kilos. Typically they have a small display, say 25 lines by 80 columns, set flat behind the keyboard; non-volatile memory; and a micro-cassette recorder, a 3.5 inch floppy drive or a hard disk for data storage. Some also have a built-in modem.

large scale integration (LSI) Integrated circuit technology involving more than 1,000 gates (that is several thousand components) per chip. This technology was developed in the 1970s, and is used for complex circuits.

laser Originally an acronym for *light amplification by stimulated emission of radiation*, lasers are valuable within the field of information technology because of their ability to direct minute light pulses very accurately. They are used within laser printers to build high-quality images; to record and read information on optical storage media such as

170

compact disks; and to transmit information along optical fibre cables.

laser printer Strictly, a printer that uses lasers to create the image to be printed, although the term is now used to describe any page printer that uses a xerographic process (such as used in normal photocopiers) for printing, some of which use other technologies such as light emitting diodes to create the image.

Laser printers work by building up a page image pixel by pixel on a light-sensitive drum. The light-sensitive drum then picks up toner and deposits it on the paper.

The earliest laser printers were expensive devices capable of speeds of several hundred pages per minute, used for high-volume printing such as mailshot letters or bank statements. More recently, much cheaper desktop printers with a speed of less than ten pages per minute have become commonplace in offices, connected to personal computers. Desktop laser printers are expensive compared to daisy wheel or matrix printers, but are quiet, flexible (they can print graphics as well as text; and can print text in a variety of styles, fonts and sizes) and capable of producing output approaching the quality of printing machines – resolutions vary from 300 dots per inch up to 1200. Maximum printing speed is typically 8 pages per minute, with some two or three times as fast.

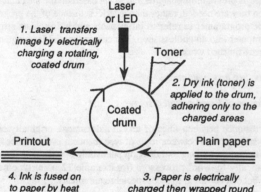

Laser or LED

1. Laser transfers image by electrically charging a rotating, coated drum

Toner

2. Dry ink (toner) is applied to the drum, adhering only to the charged areas

Coated drum

Printout

Plain paper

4. Ink is fused on to paper by heat

3. Paper is electrically charged then wrapped round the drum, picking up the ink

last-in first-out (LIFO) Describes how items in a push down stack are processed. New items are placed one by one at the top of the stack, pushing items already in the stack down one place. Items are also taken one by one from the top of the stack, which cause remaining items in the stack to 'pop' up one place again.

Compare *first-in first-out*.

late binding A concept, particularly associated with object oriented programming, where the compilation system does not resolve the question of how to execute a

171

procedure until the program is run. This means that extra features can be added to the procedure without affecting existing programs.

latency The rotational delay on a disk drive or any similar rotating storage device, in other words the time from when the read/write head is positioned over the right track until the particular sector that is required is under the head.

layer A 'slice' of a complex process that receives a defined service from the slice or layer below it, and supplies a defined service to the layer above it. The term is used particularly to refer to communications protocols, which are built up in layers of hardware and/or software, starting with the transmission and switching hardware at the bottom, right up to the processes that make use of the communications link. The reference model for open systems interconnection (OSI), defined by the International Standards Organisation (ISO), specifies seven such layers. From lowest (i.e closest to the hardware, numbered 1) to highest (7), these are physical, data link, network, transport, session, presentation, and application layers.

lazy evaluation A programming language concept based on the idea that values are only computed when they are needed, rather than in strict sequence of the program statements. This allows the programmer to refer to infinite sets and infinite data structures without putting the computer into an endless loop. The cost is that the language compiler has to keep track of part-finished work.

lazy languages See *lazy evaluation*.

LCD Abbreviation of *liquid crystal display*.

leading The distance between lines of text in a document, originally created by typesetters putting strips of lead between rows of type (hence it is pronounced 'ledding'). The term is used in word processing and desktop publishing applications. Leading is usually set up automatically by word processors (for example 12 point type normally has 14 point leading) but can be varied to meet special requirements.

learning system An area of artificial intelligence research, aimed at understanding the basic principles underlying intelligence and expressing these in systems capable of 'learning'. For example, a system might learn how to play a board game by being presented with a number of examples of positions reached in the game.

leased line A transmission line leased from a telecommunications company for private use. Leased lines, frequently referred to as private lines or private wires, normally begin and terminate on private premises, such as a business office, and cannot normally be accessed from a public network such as the telephone network.

least significant In the extreme right-hand position of a data field and thus, in a numeric field, having the least value.

LED Abbreviation of *light emitting diode*.

left justify See *justify*.

letter quality Used of printers such as daisy wheel printers that produce a character in a similar way to a normal typewriter, by means of a preformed shape.
    See also *near letter quality*.

LF Abbreviation of *line feed character*.

library routine A routine that can be called (explicitly or implicitly) into a program if needed, from a library maintained by the compiler and/or the operating system.

library subroutine See *library routine*.

library unit See *jukebox*.

LIFO Abbreviation of *last-in first-out*.

light emitting diode (LED) A transistor device that emits light when a current is present. Used individually for display lights on the fascia panel of computer systems, modems and so on, and in groups for small displays of characters that must be visible in darkness, such as on digital alarm clocks. Where there is enough light for character displays to be visible, such as on calculators or watches, *liquid crystal* technology is preferred because it uses less power. Light emitting diodes are also used to create the print image within some laser printers, and as a light source for optical fibre transmission.

light pen A device used to read printed information such as bar codes or to indicate items on a visual display screen. The former type consists of a photo-electronic device that senses the images over which it is passed and generates signals that are sent to the terminal or computer to which it is attached. The type used with a display screen usually works in a different way. The computer detects when the electron beam that scans the display tube enters the light pen (see *raster scanning*). Using the fact that there is a set scanning pattern, it can then calculate where on the screen the pen is positioned and hence what it is pointing at.
    See also *pen computer*.

limited distance modem A simplified and low cost modem, used over short distances and on private transmission lines where the stringent standards applied to equipment using the public telephone network need not be met.

line See *transmission line*.

line analyzer A device used by telecommunications engineers to 'listen in' to data communications traffic flowing along transmission lines, to help them to diagnose problems.

line control character Any character included in a transmission to control the flow of traffic on the line. This includes, for example, enquiry (ENQ) characters used for polling; start and end of text (STX and ETX) characters used to mark where messages begin and end; and end of transmission (EOT) characters.

line driver See *limited distance modem*.

line feed character (LF) A control character that tells a receiving device to advance the paper (in the case of a printer) or the cursor (in the case of a visual display terminal) one line.

line noise Extraneous signals on a transmission line, caused by interference from electrical equipment.

line printer The type of printer most commonly used at large computer installations, that prints a complete line at a time on continuous stationery. Line printers use a revolving drum or chain, or a reciprocating train to position the correct characters in a line before they hit the print ribbon.
　　Compare *character printer*; *page printer*.

line switching See *circuit switching*.

linear predictive coding A technique used to produce synthetic speech. It produces good results but requires considerable processing power. It involves sampling and recording a voice digitally many times per second. From these samples, software predicts what the voice will sound like during the time when it was not sampled, thus rendering a continuous voice.

linear programming A form of mathematical programming used particularly to solve allocation problems. Linear programming can be used to maximise or minimise a function which is the sum of multiples of several variables, subject to constraints on those variables that can be defined in a similar way.

lines of code See *source lines of code*.

lines per minute (LPM) The normal measure of the speed of line printers.

link See *communications link*.

link access protocol (LAP-B) A class of protocols for controlling data transmission over a communications link, including correcting transmission errors. Some specific protocols have been defined, identified by an alphabetic code, and notably LAP-B and LAP-M which have been incorporated into a number of communications products, such as synchronous modems and local area networks.

link layer See *data link layer*.

linked list A method of organising a collection of related items, such as a string of detail records belonging to a master record. In this case, the master record is linked to the first item (by means of an address field or pointer), then each item is linked to the next in the chain (or list), and the final item is linked back to the master record. An item can be added to or removed from the chain at any point simply by changing the link in the preceding item to point either to the new item or to the item beyond the deleted one.

linker A piece of software that forms part of a compilation system. Its job is to interconnect the various components of a program assembled in a preceding compilation phase.

lips Abbreviation of logical inferences per second, a measure of the performance of a knowledge-based system.
    Compare *mips*.

liquid crystal display (LCD) A technology used for display screens. It works by applying a voltage to dense liquid held between two sheets of glass. It requires very little power and space, so is ideal for the screens of battery-powered devices such as portable computers. Two LCD technologies are used: *supertwist* (or passive matrix) technology does not create as sharp images as the more expensive *active matrix*, and also suffers from the problem of shadowing, sometimes called ghosting. Compared with CRT (cathode ray tube) technology, the dominant technology for display screens, LCD screens have the advantages of perfect geometry and are nearly free from electromagnetic emissions, but they are more expensive and less bright.

LISP Abbreviation of LISt Processing, a high-level programming language designed for manipulating non-numeric information, such as words or statements. It is one of the main languages used both for research into artificial intelligence and for its commercial application.

literal A constant that is held within a program instruction or statement, rather than being declared separately and referenced by means of its identifier.

liveware The people who are involved in operating or using an information system. Humorous extension from hardware and software.

load and go Where a source program is run immediately it has been compiled, rather than being stored as an object program that can be run as a separate step.

load point See *beginning of tape marker*.

load testing See *saturation testing*.

loader An operating software routine that loads object programs into memory ready to be run.

LOC Abbreviation of *source lines of code*.

local (1) On the same site. Contrast with *remote*, meaning on another site.
    (2) Extending over one site only, or over a limited area – see *local area network*.
    (3) In programming, applying only within the current routine – see *local variable*.

local area network (LAN) A type of switched communications network operating over a limited distance, such as across a site or within a building, and carrying non-voice communications traffic such as data and text, and sometimes image and video also. Local area networks usually connect a number of relatively small devices such as personal computers, minicomputers and visual display terminals, enabling them to exchange

Schematic of typical small local area network installation

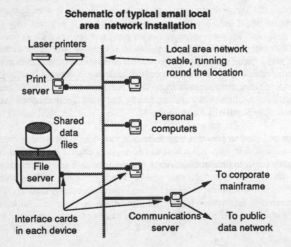

176

information and share resources with great flexibility. Some types can support a mix of different types of device, others are designed only for a particular manufacturer's equipment.

LANs vary in transmission medium and in topology. Optical fibres, coaxial cable or twisted pair (telephone) wiring may be used as transmission medium, and topologies include bus, star and ring arrangements. Speed of transmission also varies, according to the equipment supported by the network. Three main types of protocol are used – polling, token passing and contention. The most popular version of the last is *carrier sense multiple access*, or CSMA for short. See also *Ethernet* and *Cambridge Ring* – two specific LAN technologies.

Also used in a more general sense to mean the site or building network, as opposed to the wide area network that may interconnect sites and buildings.

local bus An extra bus within a computer that carries data between the main bus and one particular component, such as the display screen (often referred to as *local bus video*).

local exchange See *exchange*.

local line The transmission line connecting a device – telephone, data terminal, computer – to its local exchange.
Compare *trunk line*.

local network See *site network*.

local variable A variable that is only accessible to the element of a program in which it is declared.
Compare *global variable*.

locality (of data) Used to describe how regular is the pattern of occurrence of transactions of different types, particularly when dealing with transaction processing applications involving a number of different sites. With high locality of data, most of the transactions of a particular type originate at the same point, and this makes it advantageous to site the files those transactions affect at that point. If locality of data is low, by contrast, transactions will often need to be forwarded from the point where they originate to the point where the relevant data files are held. Where this is the case, centralised data files are likely to give better performance.

localise Convert to conform to local requirements, normally used of software packages. For example, some personal computer software packages are supplied in a single version that can easily be localised to support the language and other conventions used in different European countries.

location See *memory location*.

177

location independence Used of data files or other resources distributed across a number of sites, meaning that they can be used in the same way regardless of where they are located.

log off Relinquish a connection with a service provided by a computer system, by sending a closing message of some kind. The computer will acknowledge the closing message, often indicating what resources (time, processor activity, etc.) have been used during the session that has just ended.

log on Establish a connection with a service provided by a computer system, prior to using it. Logging on normally involves sending messages from the terminal to the computer system to specify which service is required and to identify the user at the terminal. A password or an account number may also have to be entered to confirm the identity of the user. See *password protection*.

logic The way a component of a computer system solves the particular problem it is addressing. Used both of hardware (see *hard-wired logic*) and software (see *program logic*).

See also *fuzzy logic*.

logic board The printed circuit board in a small computer that carries the main system components such as processor and memory chips.

logic bomb Programming code 'hidden' within a program, designed to cause damage, such as by deleting or corrupting data files, when given circumstances arise. Logic bombs are occasionally left behind by disgruntled programmers looking for revenge on their former employer.

See also *virus*.

logic circuit See *logic element*.

logic element A component of an electronic circuit that performs a logical operation. Digital computer circuitry is based on boolean algebra, which defines six basic types of logical operation. Both memory and processing circuits are built up from combinations of logic elements (or gates) carrying out these basic operations. They are:

1) AND circuit. A circuit with two or more inputs and one output whose output is high if and only if all the inputs are high.

2) Inverter (NOT gate). A circuit with one input whose output is high if the input is low and vice versa.

3) NAND circuit. A circuit with two or more inputs and one output whose output is high if any one or more of the inputs is low, and low if all the inputs are high.

4) NOR circuit. A circuit with two or more inputs and one output whose output is high if and only if all the inputs are low.

178

5) OR circuit. A circuit with two or more inputs and one output whose output is high if any one or more of the inputs are high.

6) Exclusive OR circuit. A circuit with two or more inputs and one output whose output is high if any one or more of the inputs (but not all) are low.

To translate into terms appropriate for digital logic, for low read zero, and for high read one.

logic programming Programming by expressing facts, relationships and rules in logical statements. It is used for applications that manipulate symbols (such as words or statements) in an intelligent way, such as natural language processing or expert systems. The most widely used logic programming language is Prolog.

logic seeking Describes a printer that modifies its operation based on the information sent for it to print. For example, it checks the length of the next line to be printed and decides whether it would be quicker to print it from right to left or vice versa, depending on where the print head will be after printing the preceding line.

logical Generally used to mean a resource – a channel, a file, a unit, a memory location – which can be treated by a program as if it were an actual resource, but where a particular physical resource is assigned only when the need arises. For example, a program processes a logical file which it calls 'Input file' (or any other convenient name). At the beginning of the program, it will issue an Open command, specifying both the logical file name ('Input File' or whatever) and identifying details of the actual file required (the physical file). This will cause the operating system to locate the physical file on a particular peripheral unit. Subsequently, it will interpret all references by the program to the logical file as applying to the particular physical file.

By using logical rather than physical references, programs can be executed in a variety of different circumstances without needing to be amended, because decisions on physical resources can be made at run time by operating software or by computer operators.

See also *logical record*.

Compare *physical*.

logical AND See *AND operation*.

logical operator An operator that defines a logical relationship between two conditions, such as AND, OR or NOT.

logical OR See *OR operation*.

logical record A record from a data file as processed by an applications program. Contrast with *physical record* or *block*, which is the form in which records are held on storage devices. Commands such as 'put record' and 'get record' are used by applications·

179

programs to manipulate logical records. Outgoing logical records are packed into blocks by file or database management software, and similarly incoming logical records are unpacked and passed across to applications programs individually.

logical shift An operation which causes the bits in one or more words to shift a specified number of bits in either direction. Bits shifted out of either end reappear at the opposite end. In other words, the word or words operated on act as if they were circular. Also known as *circular shift*.
Compare *arithmetic shift*.

Logo An interactive programming language, designed for educational purposes by a US computer scientist called Seymour Papert.

long word An unusually long unit of computer memory, used by particular instructions that need extra storage space for operands. For example, a computer system with its memory organised as 16-bit words may use 32-bit long words as operands in high-precision arithmetic instructions.

longitudinal redundancy check (LRC) A parity check that is applied to a series of characters, such as those in a transmission block. Each bit position in successive characters is treated as a unit, and a parity bit is generated. Together these parity bits form an extra character that is appended to the original series of characters.
Compare *cyclic redundancy check*.

lookup table An array of data fields organised so that individual items can be taken from it when needed using a variable (the subscript) to identify the item required. For example, a program might have a lookup table of error messages and use an error code as subscript.

loop A sequence of program steps that is repeated a controlled number of times. Programmers use control structures such as REPEAT or DO WHILE statements for the purpose. Also used in the phrase 'in a loop', to describe a program error in which a loop occurs accidentally or where control fails so that the loop never completes.

low level language See *assembly language*.

low-level vision See *machine vision*.

low order See *least significant*.

low speed modem A modem operating at speeds up to 1200 bits per second.

LPM Abbreviation of *lines per minute*.

180

LRC Abbreviation of *longitudinal redundancy check*.

LSI Abbreviation of *large scale integration*.

luggable Describes a personal computer that is portable but not as small and light as a laptop. Normally mains-powered.

luminance The brightness of a visual display screen or of lighting in a room, usually expressed in candelas per square metre.

LZW algorithm A data compression algorithm (taking its name from its designers Lempel-Ziv-Welch) that is a refinement of *Huffmann encoding*. As well as encoding individual symbols in the data it also encodes sequences of symbols. It can shrink coded text by 50 to 60 per cent, monochrome bit-mapped graphics by 80 per cent, and program code by 40 per cent or less.

Compare *Huffmann encoding*.

M

M Abbreviation of *mega-*.

machine address See *absolute address*.

machine code Program instructions in the form in which they are executed by a processor.

machine intelligence See *artificial intelligence*.

machine language A programming language that corresponds closely to machine code. These so-called first-generation languages were introduced in the early stages of computing, from the mid-1940s onwards.
  Compare *assembly language*; *high-level language*.

machine learning See *learning system*.

machine readable Information recorded in such a way that it can be interpreted directly by a device for entry to a computer. Bar codes on food packages and magnetic ink characters on cheques are examples.

machine vision A branch of artificial intelligence work aimed at building machines capable of simulating human vision. It can be divided into two main categories:
  (1) low-level vision uses simple numerical algorithms to carry out repetitive operations on large images;
  (2) scene understanding, or high-level vision, involves sophisticated decision-making to search, match and verify elements making up the image.

macro A single statement or command that automatically enacts a predefined series of statements or commands. In programming, macro statements are defined for sequences of statements that are used frequently. The definitions of some macros, such as those used to call operating system routines, are usually built into the programming language, and programmers can also define their own macros, by specifying the sequence of statements to be generated and the name of the macro statement they wish to use to generate them. Macro commands can also be set up on personal computers, to replay a particular sequence of commands or operations automatically.

macro flowchart A flowchart showing a system or program in overview, rather than in the full detail necessary to implement it.

macro generator A program that reads in a source program and expands the macro statements it contains.

macro statement See *macro*.

magnetic bubble memory See *bubble memory*.

magnetic card A storage medium consisting of cards with a magnetisable surface, now obsolete. Cards were identified by notches along their upper edge. They were held in magazines from which they could be selected and transported past read/write heads before being returned to the magazine.

magnetic disk See *disk*.

magnetic drum A storage device consisting of a cylinder coated with magnetisable material, rotating continuously past a series of read/write heads, each of which coincides with a track on the surface of the cylinder. Drums originally had a performance advantage over disk storage devices, because it was not necessary to position the read/write head over a track before reading or writing data. Drums have since been overtaken by developments in disk technology.

magnetic ink character recognition (MICR) The reading, by electronic means, of characters printed in a special magnetic ink. This technology is used to record account and cheque numbers on cheques throughout the world. The characters are printed in a highly stylised form using ink impregnated with magnetisable particles. For reading, the ink is magnetised, then the characters are passed across a read head. Each character is recognised by its characteristic shape, and converted into digital codes that are used to identify characters within a computer system.
Compare *optical character recognition*.

magnetic resonance imaging (MRI) A technique used to create an image of the internals of the human body. The body is placed in a magnetic field. The pattern of ions round the body is metered and this is used by the computer to build an image of the

183

tissues that is displayed on a screen. The operator of the device can also use it to strip away layers of tissue, thus revealing more detail than is visible on a normal X-ray picture. The technology on which these imaging devices are based is known as nuclear magnetic resonance (NMR).

magnetic tape See *tape*.

magnetic tape label See *header label*.

mailbox See *electronic mailbox*.

mailmerge Produce a series of versions of a basic document (usually a letter) incorporating into each information taken from successive records in a data file. Mailmerge is used to incorporate names and addresses into a mailshot letter, and may also be used to include different sections of text in the letter, based on codes held in the data file along with the names and addresses.

main memory See *memory*.

mainframe computer A general term for a large centralised computer system, which were originally built on a large frame or chassis. Such a system is capable of handling a more varied workload than its cousin the minicomputer, and especially mixed loads of batch processing, teleprocessing and timesharing, in other words the typical workload of the heyday of centralised data processing in the late 1970s and early 1980s. At first, there were distinct hardware differences between mainframes and minicomputers, but these have largely disappeared, and the significant remaining differences are in operating software. Typically, a mainframe has the more sophisticated job scheduling and job control facilities that a varied workload demands, at the cost of greater operating complexity.

maintenance See *program maintenance*.

maintenance contract A contract between a user organisation and a computer supplier or a computer maintenance company, specifying the terms under which a computer system will be maintained. As well as financial arrangements, this will normally specify how quickly maintenance engineers will respond to reports of failures, what the level of availability of the system should be, and when routine maintenance will be carried out.

man-machine interface See *human-computer interaction*.

managed data network services (MDNS) A complete data network service, including operation, maintenance and planning, rather than data transmission alone as is available from common carriers such as British Telecom or Mercury. Managed data network

184

services are a category of value-added network services (VANS), requiring a special licence from the regulatory body (Oftel in the UK).

management information system (MIS) This term, often abbreviated to MIS, was first coined when early data processing systems outgrew their original role of controlling basic business processes, and began to be adapted to meet managers' information needs. Data processing departments changed their names to 'MIS departments' to reflect this change of emphasis. The term is used in a general sense to mean the applications developed and run by the central systems group, as opposed to those (such as personal computers and word processors) controlled by end-users. It is also used in a narrower sense to mean an application specifically designed to supply managers with information.

management workstation See *workstation*.

manual input Entry of data by a human being, such as via a keyboard.

manual operation An operation carried out without the help of a computer system, such as fetching a document from out of a filing cabinet or picking goods in a warehouse.

manufacturer lease A type of lease offered by computer manaufacturers which has the same advantages as an operating lease but with restrictions aimed at tying the lessee into the manufacturer's own equipment.

manufacturing automation protocol (MAP) A set of supplier-independent protocols for data communication in and with the factory, first proposed by General Motors. The MAP protocols are based on the OSI Reference Model defined by ISO, and have received the support of a number of large manufacturing companies.

manufacturing resource planning (MRP) A complex computer-based production-planning technique mainly used by manufacturers of complex products, such as engineering companies. Often referred to as MRP or MRP II – to distinguish it from the earlier material requirements planning, also abbreviated to MRP.

MAP Abbreviation of *manufacturing automation protocol*.

mark reading See *optical mark reading*.

mark sensing See *optical mark reading*.

mask As a verb, select a number of bits within a bit pattern. This is done by superimposing one bit pattern over another. The former, known as the mask, has bits set in it to indicate which bits in the latter are to be preserved; the remaining bits are zeroised.

185

mass storage Storage devices designed for large volumes of data, such as disks and tapes, and that are connected to a computer as peripherals. Contrast with a computer's *internal memory* (often referred to as RAM, for *random access memory*) whose capacity is much more limited, and which can be accessed directly by the processor, whereas data is transferred to and from mass storage peripherals by means of input/output commands. Also known as *backing storage, bulk storage, filestore* and *external memory*.

massively parallel processing (MPP) Parallel processing involving such a large number of processors that they cannot share a single memory efficiently and instead use a distributed memory architecture. This in turn places heavy demands on the controlling software which, in a database application for example, has to apply locks to make sure that any two records are not being changed at the same time.

master console See *operator console*.

master file A data file containing information which changes relatively infrequently and which is used as source of reference data by applications programs, such as a file identifying customers or products.
  Compare *transaction file*.

master record See *master file*.

master-slave A relationship between communicating devices, in which the master (normally the central computer system) determines when the slaves (the remote terminals) may and may not send or receive messages. Master-slave relationships are normal in teleprocessing networks, and are reflected in the communications protocols, such as IBM's binary synchronous communications (BSC), used in those networks. As remote terminals are replaced by more powerful devices, such as departmental and personal computers, master-slave relationships and protocols are gradually giving way to peer-to-peer relationships, in which any device may initiate an exchange of messages.
  Compare *client-server*.

master terminal A designated terminal within a network of terminals which has special supervisory powers, such as to reconfigure the network or to change the authorisation levels of terminal users.

material requirements planning See *manufacturing resource planning*.

maths coprocessor See *floating point unit*.

matrix matching The original technique used for optical character recognition. The unknown character is scanned into memory and then compared pixel by pixel with a library of reference standards. Once a match is found, the character is converted to the

186

equivalent character code. This technique works well only for documents limited to a few typefaces or point sizes. Where many variations occur, an alternative technique called feature extraction is more effective.

matrix printer A printer that forms each character out of a matrix of tiny dots, produced by striking the print ribbon selectively with a set of tiny needles. Matrix printers are cheap and capable of operating at speeds up to 500 characters per second. Some devices make multiple passes across the paper, distributing the imprint so as to make a better letter image. Users of these devices can choose between draft quality output at high speed, or what is sometimes called near letter quality output at slower speeds. They are also capable of printing simple graphics.

Mb Abbreviation of *megabyte*.

Mbit Abbreviation of *megabit*.

MCA Abbreviation of *micro channel architecture*.

MCGA Abbreviation of *multi colour graphics array*.

MDA Abbreviation of *monochrome display adaptor*.

MDNS Abbreviation of *managed data network services*.

mean time between failures (MTBF) The average time for which a piece of equipment works without failing. More precisely, the ratio of the total time in a given period to the number of failures in the period. This is a standard measure of equipment reliability, often coupled with mean time to repair (MTTR).

mean time to repair (MTTR) The average time taken to repair a failure in a particular piece of equipment. More precisely, the ratio of time spent in corrective maintenance of equipment, in a given time period, to the number of failures of that equipment. This is a standard measure of equipment reliability, often coupled with mean time between failures (MTBF).

medical imaging The creation and display of images for medical purposes, such as in body scanners. The computer is used for such purposes as to enhance images to make them easier to read; to build up a complete picture from a series of 'slices'; and to manipulate images to look at them from different directions.

See also *computerised tomography*; *digital vascular imaging*; *magnetic resonance imaging*.

medium scale integration Integrated circuit technology involving up to 1,000 logic elements (that is several thousand components) per chip. This technology was developed in the late 1960s and early 1970s, and is used for circuits of modest complexity, such as a counter.

medium speed modem A modem operating at speeds between 1200 and 4800 bits per second.

mega- (M) A prefix meaning millions of (x 10^6), as in megabit or megabyte. Often abbreviated to M, as in Mbit.

megabit (Mbit) A unit of data volume. In megabits per second, a measure of the speed of transmission of digital data. For example, an Ethernet local area network operates at a speed of 10 megabits per second (abbreviated to 10Mbit/s or 10Mbps).

megabyte (Mb) A measure of storage volume, meaning 1 million bytes. It is often abbreviated to Mb or Mbyte and commonly applied to computer memory and disk storage. Like kilobyte, it means the nearest power of 2 when used to measure the capacity of computer memory, since digital computers use binary numbers internally. This is 2^{20} or 1,048,576. Thus a two megabyte computer has 2,048 kilobytes of memory, which equates to 2,097,152 bytes.

megahertz (MHz) A measure of frequency and hence of processing or transmission speed, meaning one million cycles per second.

MegaStream British Telecom's trade name for its point-to-point digital transmission services with speeds in excess of 2 megabits per second (Mbit/s). MegaStream services can be used either to carry telephone conversations (up to 30 simultaneously) or to carry data, or a combination of the two. MegaStream services are typically used to interconnect mainframe computers, local area networks or private telephone exchanges.

memory (1) A general term for the devices used by computer systems to store information – both data and programs – before, while and after it is processed. This can be seen as a hierarchy. As you proceed down the hierarchy, speed of access and cost per byte of storage both decrease. At the top of the hierarchy is the *read-only memory* (ROM) used within a computer to hold frequently used parts of the operating software; this is followed by the computer's main (also known as *primary*) internal memory, in the form of RAM (*random access memory*) chips; then by a whole range of mass storage devices of varying characteristics, sometimes referred to as *external* or *secondary memory*.

(2) More particularly, the component of a computer in which programs and data are held while they are being executed and processed, respectively. Sometimes called main or internal memory or storage for contrast with *mass storage*, from which programs and data must be transferred into memory before they can be executed or processed.

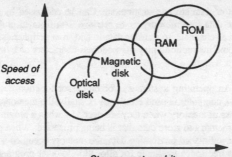

memory capacity The amount of memory installed in a particular computer, usually expressed in thousands (suffix 'K' for kilo-) or millions (suffix 'M' for mega-) of bytes or words. Thus a 256K computer has 256,000 bytes of memory (approximately – see *kilo-*).

memory card A card containing semiconductor memory that can be plugged into small computers. It has the same size and appearance as a credit card, and is used particularly with portable computers, often to transfer data recorded on the portable computer to a larger computer for later processing. Also known as *silicon drive*.

memory compaction A process designed to make available as large a contiguous area of memory as possible. Both in the memory of computer systems and on direct access storage devices such as disks, the available area is allocated for use as and when required, and is released when no longer needed. In the course of time, this can mean that the area becomes fragmented into small pieces, since it often becomes necessary to split up areas to meet demands for pieces of varying size. Eventually, the available area has to be re-assembled, moving allocated areas if necessary, and this process is known as compaction.

memory cycle The time a processor takes to retrieve a word of data from memory.
　　See also *control unit*; *cycle time*.

memory dump A printed representation of some or all of the contents of computer memory. A memory dump may be produced automatically by the operating system when a program fails or as a precaution against future failure, or on request by a programmer or computer operator. The programmer uses it to reconstruct the preceding sequence of events, so that programming errors or software malfunctions can be identified.

memory fragmentation Where the available space in an area of memory is in a large number of separate, small pieces. This situation often arises after a time in an area of memory that is allocated dynamically.
　　See also *memory compaction*; *memory management*.

memory location A unit of data storage in memory. Can be addressed by program instructions, either to store data into memory or to retrieve data from it; and also by input/output hardware such as channels, to transfer data to and from peripherals. Usually addressed in words, but some instructions may also address characters or bytes within words.

memory management An operating system and/or processor function designed to ensure that the memory of a computer is used efficiently. Usually, the memory management routine allocates blocks of memory when they are needed – when a program is first loaded or when it asks for memory to store data that is being processed. When programs are deleted or when they have finished processing data, the memory occupied is released and the routine marks it as free space available for allocation to other programs. In the course of time, this free space can become fragmented, because the routine cannot find blocks of memory of exactly the size required and has to keep splitting up larger blocks. To solve this problem, either periodically or when it cannot find a single block large enough to meet demands for memory, the memory management routine will compact the free space by moving the blocks of allocated memory around and combining adjacent blocks of free space together.

memory protection Mechanisms to prevent accidental interference between different programs running in a multiprogramming computer system. Each program is allocated a particular area of memory to use. Should any of its instructions generate a memory address outside this area, for example by using an incorrect indirect address, memory protection hardware intercepts and prevents access to that memory location and informs the operating system, via the processor, which in turn stops the program and informs the operator.

memory-resident Held continuously in the memory of a computer system, rather than being loaded and kept in memory only when needed. Routines that are used heavily or that must be available with minimum delay, such as parts of the operating system, are memory-resident so that they can be executed without the extra processing incurred by loading them from direct access storage such as disk.

menu A list of the options available from a computer system at a particular stage of processing, displayed on the screen. To select an option, the user either types one of the characters identifying each option or moves the cursor to the chosen option and selects it in some way, such as by pressing the Return key or the mouse button.
    See also *pop-up menu*; *pull-down menu*.

menu-driven Of a program that is controlled by selecting options from menus rather than, for example, by keying in a series of commands.
    Compare *command-driven*.

menu selection See *menu*.

merge Combine two or more sequential files to create a single data file whose records are in the same sequence as the originals.

Compare *sort*.

message In general, a collection of data communicated from one point in a system to another in order to convey a particular meaning – for example, an error message displayed on a screen. Used more specifically in the context of data transmission and data networks to mean a collection of data to be transmitted as a unit.

See also *packet*.

message router See *router*.

message switching A technique used to transmit messages (usually consisting of text) from one point to another. It consists of accepting a message, storing it until a means of forwarding it (such as an outgoing line) is available, and then transmitting it onwards to its destination, usually via a number of such stages. Forwarding may also be delayed in order to take advantage of off-peak transmission charges. No direct connection is set up between caller and addressee, as it is in circuit switching. Also known as *store-and-forward switching*.

Compare *circuit switching*; *packet switching*.

message transfer See *data transfer*.

metadata Data which describes data. Metadata helps a database administrator and applications programmers to carry out their respective tasks, and is often held in a data dictionary – a special software package designed to record and manage metadata. Metadata comes in three varieties:

(1) Semantic metadata describes the meaning of data.

(2) Physical metadata describes how the data is represented in storage – field size, field type, frequency of occurrence and use.

(3) Usage metadata describes what the data is used for and by whom.

method In the terminology used to describe object-oriented programming, the things that can be done to a particular object. Thus the methods applicable to an object 'loan' might be 'authorise', 'extend', 'redeem' and 'write off'.

metrics See *software metrics*.

metropolitan area network A network extending across more than one site but with a limited range, such as across a metropolitan area or a university campus. Thus, in terms of range, a metropolitan area network falls between a local area and a wide area network.

MFM Abbreviation of *modified frequency modulation*.

MHS Abbreviation of *X.400*.

MHz Abbreviation of *megahertz*.

MICR Abbreviation of *magnetic ink character recognition*.

micro See *microcomputer*.

micro channel architecture (MCA) A new type of high-performance channel for personal computers, designed by IBM for its PS/2 range of personal computers. IBM's powerful position in the market encourages other suppliers to conform to designs such as this for their own products. An alternative standard is being promoted by a number of suppliers, called extended industry-standard architecture (EISA).

microcode See *firmware*.

microcom networking protocol (MNP) A set of protocols for controlling data transmission by a modem over a communications link, including correcting transmission errors and compressing data. There are a number of versions of this protocol of varying sophistication, identified by level numbers running from 1 up to 7. MNP protocols are supported by a number of different modems and by bulletin boards.

microcomputer A small computer system, usually with a single-chip processor and assembled on a single printed circuit board. The microcomputer is best known in the form of personal computers, but it is also incorporated in other forms of small-scale equipment of limited power, such as games machines, intelligent cash registers, supermarket checkout terminals, hole-in-the-wall banking terminals, and so on.

microelectronics The technique of fabricating a large number of electronic components on small quantities of semiconductor materials such as silicon, together with the electrical connections between them. The resulting product is known as an integrated circuit, or more colloquially, a silicon chip. At present, it is possible to manufacture in excess of 100,000 circuit elements on a single chip a few mm. across – a technology known as very large scale integration, or VLSI.

microform A general term for documents held in a miniaturised form using photographic techniques, such as on microfiche.

micrographics A group of technologies that uses photographic techniques to produce miniaturised images of documents or drawings. It includes microfilm and microfiche storage systems, retrieval from which can be computer controlled, and also computer output microfilm (COM).

microjustification The process whereby minute spacing increments are used to measure and adjust spacing between characters and words in a document. Good microjustification, supported by advanced word processing and desktop publishing applications, results in more readable and attractive printed output, similar to typeset material.

micronet A local area network designed to interconnect personal computers. Micronets are relatively cheap to install and usually support only a limited range of equipment.

microprocessor A processor on a single integrated circuit or chip. Most microprocessor chips consist of at least the following functional units (see figure):

- a control unit to sequence processing and to decode program instructions;
- an arithmetic/logic unit to perform the arithmetic, logical and other operations necessary to carry out the program instructions;
- registers and accumulators to hold data while instructions are carried out;
- an address buffer to hold the address of the next instruction;
- input/output buffers to hold data flowing to and from the processor.

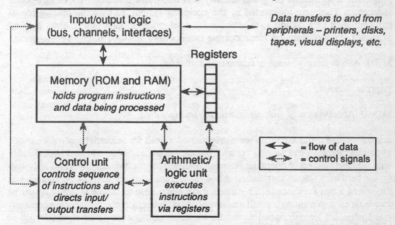

The first microprocessor, the Intel 4004, was produced in 1971 for incorporation in a desk calculator, and had a word length of only 4 bits. Microprocessors with an 8-bit word (often called 8-bit microprocessors) powered the first generation of personal computers, typified by the Apple II. By today's standards they have limited power and are only able to address a limited amount of memory. 16-bit microprocessors can address hundreds of kilobytes of main memory, and power machines such as the original IBM PC. The latest microprocessors have a word length of 32 bits and a sophisticated instruction set, and power personal computers and scientific workstations, as well as a wide range of other small-scale equipment. As well as the components listed above, modern microprocessor chips often include units for floating point arithmetic and memory management, and separate cache memories for data and instructions.

The leading designers and manufacturers of microprocessor chips are US companies Motorola and Intel, both of whose products are identified by numeric product codes. Intel's current range of microprocessors have codes beginning with 8, including the widely-used 8088 and the more recent 83086. Motorola's successful chips include the 68000 and the more recent 68030.

microwave transmission A transmission technology using radio waves, sometimes used in communications networks for short point-to-point connections, such as between adjacent buildings or across natural obstacles.

middleware Software that acts as an intermediary between an operating system and applications programs. Its purpose is to make the task of applications programs, and of the programmers that develop them, easier. Middleware performs frequently used complex tasks and tasks that require co-ordination between a number of applications programs. Applications programs usually call middleware routines by using special statements. These are interpreted by the operating system, or by some other piece of software such as a compiler, to activate the appropriate middleware routine. Typical examples of middleware are teleprocessing monitors and database management systems.

MIDI Abbreviation of *musical instrument digital interface*.

migrate See *port*.

MIMD Abbreviation of *multiple instruction multiple data*.

minicomputer Minicomputers were originally designed for industrial applications such as plant monitoring and process control, but their application has widened considerably with the development of operating software designed for commercial applications. Initially, the hardware feature that distinguished them from other commercial computers was the use of a bus architecture. This made them better able to handle a large number of terminals each generating a small amount of traffic intermittently, where other commercial computers were designed to handle the continuous, steady load characteristic of batch data processing. This clearcut hardware distinction has now disappeared, and the term minicomputer is used to describe medium-sized machines, usually serving a department or a small organisation. This is in distinction to, at one extreme, a mainframe computer serving a large organisation or, at the other, a personal computer serving an individual.

Minitel Terminals distributed by the French PTT that are used to access an electronic telephone directory service and other videotex services. It is the largest videotex network in the world, with over 4 million terminals installed at the beginning of 1989.

mips Abbreviation of millions of instructions per second – a crude measure of the performance of a computer processor.
Compare *FLOPS*.

MIS Abbreviation of *management information system*.

MIS department See *information systems department*.

MM Abbreviation of *multimode*.

mnemonic codes Codes used in assembly languages to represent instructions. They are short (usually three or four letters) abbreviations of the machine code operations, for example MVC for 'move character' or LDA for 'load register A'.

MNP Abbreviation of *microcom networking protocol*.

mobile data communications Communication of data direct to and from mobile terminals, such as hand-held terminals consisting of a keyboard and small screen or laptop personal computers. Much as for mobile telephone services, suppliers of these services set up base stations distributed across the geographical area they wish to cover. Messages are exchanged between base station and terminal by radio and typically are transmitted between base stations by means of a conventional packet switching network. Outgoing messages that cannot be delivered immediately may be stored at base stations until the destination terminal moves within range.
Compare *packet radio*.

modal dialogue A dialogue between a user and an applications program that the user must complete before being allowed to do anything else, such as specifying which files are to be opened for processing.
See also *dialogue management*.
Compare *modeless dialogue*.

modeless dialogue A dialogue between a user and an applications program that the user can interrupt if he or she wishes, and come back to later, for example a 'find and replace' operation in a word processing application.
See also *dialogue management*.
Compare *modal dialogue*.

modem A piece of equipment used to link a digital device such as a computer or terminal to an analog telephone line. The term is a contraction of modulator-demodulator, and one of the main functions of the device is to modulate an outgoing stream of digital data bits so that it is compatible with telephone networks designed to handle analog speech traffic, and to reverse the process with an incoming bit stream.

Compare *network interface unit*.

modified frequency modulation (MFM) The standard used by IBM-compatible personal computers to initialise disks.

Compare *group-coded recording*.

modifier A data field that is added to another so that this can be used to access successively the items in an array. Normally, a modifier is used within a loop and is increased on each cycle through the loop.

Modula 2 A programming language devised in 1980 by Niklaus Wirth, the inventor of the Pascal language. It is a very rigorous language, demanding a highly disciplined approach from the programmer, with the object of producing more reliable software. It is widely used within the academic world in Europe but so far has gained little acceptance within business.

modular programming A technique intended to simplify the task of developing large programs or systems. Before coding starts, programs are divided up into a number of self-contained logical sections or modules. Each of these is developed and tested separately (using a test harness which supplies test data and simulates the other modules with which it interacts), then the modules are progressively assembled together until the complete program has been built up.

modulation The process used by a modem to imprint a digital signal on to a carrier wave, so that it can be transmitted over networks designed for analog traffic, such as the public telephone network.

See also *pulse code modulation*.

module A functionally separate section of a program, that can be tested independently of other modules of the program.

See also *modular programming*.

moiré A fine multicoloured pattern made up of concentric ovals. It appears, for example, on large areas of light grey on display screens, sometimes disappearing when the image brightens.

monitor A visual display screen. Usually means the screen alone, as distinct from the processor and other logic which drives it.

monochrome Consisting of black (or some other colour) and white only. Used of visual display screens and images. Contrast with *grey scale*, which includes shades of black-ness, and also, of course, with colour displays and images.

monochrome display adaptor (MDA) A basic standard for the display of information on the screen of a personal computer, and also a plug-in expansion card that implements the standard. It was introduced along with IBM's original PC in 1981, and provides for text characters to be displayed on a monochrome screen, constructed from a 9 x 14 matrix of dots.

monomode See *single mode*.

monospaced Used to describe a font or a line of text where every character has the same width.
Compare *proportional spacing*.

Moore's Law A law formulated in the 1960s by Gordon Moore, one of the founders of chip manufacturer Intel. As originally formulated, it stated that the number of circuit elements per integrated circuit would double every two years. In 1976 he revised the law to reflect the quickening pace of development, and it now holds that the number doubles every 18 months, or in other terms increases at a rate of about 60 per cent per year.

mosaic graphics See *alphamosaic*.

most significant In the extreme left-hand position of a data field and thus, in a numeric field, having the greatest value.

motherboard The printed circuit board on which is mounted the processor, memory and input/output logic of a personal computer or similar small system. Where these are provided, it will also include the expansion slots in which plug-in expansion cards, or daughterboards, can be mounted. Also known as *backplane*.

mouse A device used to control a personal computer or a graphics workstation. It consists of a ball set under a small case. The ball rolls as the mouse is moved around on a flat surface, and those movements are reflected in the movements of a pointer or cursor on the display screen. The mouse also has one or more buttons that can be pressed to send a signal to the computer. To select an item from a menu, for example, the mouse is moved to position the pointer over a menu option, then a button is pressed to select that option. The mouse is also used to enter non-character information, such as drawings.
See also *graphical user interface*.
Compare *tracker ball*.
More detail *drag*.

MPEG algorithm An algorithm designed by the MPEG (Motion Picture Experts

Group) to be used to compress live video signals so that they can be stored economically on a hard disk or CD-ROM. The aim is to compress sound and vision information by a factor of more than 100, both by compressing individual frames and by considering similarities between successive frames.

Compare *JPEG algorithm.*

MPP Abbreviation of *massively parallel processing.*

MPX Abbreviation of *multiplexor.*

MRI Abbreviation of *magnetic resonance imaging.*

MRP Abbreviation of *manufacturing resource planning.*

MS-DOS The most widely used operating system for 16-bit personal computers. It was developed by Microsoft in cooperation with IBM. An adapted version, PC-DOS, is available on IBM's own PC.

MTBF Abbreviation of *mean time between failures.*

MTTR Abbreviation of *mean time to repair.*

multi-access computing See *timesharing.*

multi colour graphics array (MCGA) A standard for the display of information on the screen of a personal computer, introduced by IBM for the smaller machines in its PS/2 range. Text characters are constructed from an 8 x 16 matrix of dots. There are two graphics modes: medium (256 colours or 64 grey shades, with a resolution of 320 x 200 pixels) and high (two colours or two grey shades, 640 x 480).

multi-reel file A magnetic tape file that occupies more than one tape reel.

multi-user Of computer systems, able to be used by more than one person at the same time. This is achieved mainly via the operating system, which shares out the system resources, such as processor and memory, among the programs being run by different users.

multidrop See *multipoint.*

multifunction workstation See *workstation.*

multimedia See *interactive multimedia.*

multimode (MM) Used of optical fibre that allows several rays of light to propagate along the fibre. The core of a multimode fibre is relatively large – ranging from 50 to 85 microns in diameter – and can accept light signals from cheap components such as light emitting diodes (LEDs). Contrast with *single mode* (SM) fibre, which is more expensive but has better performance.

multiple instruction multiple data (MIMD) A type of parallel processing machine that farms out different parts of the program it is running, dealing with different bits of data, on to different processors.
Compare *single instruction multiple data*.

multiplexer See *multiplexor*.

multiplexing Techniques that permit the sharing of communications links among a number of low-speed user devices, such as data terminals. Frequency division multiplexing (FDM) lets a number of low-speed devices share a link by dividing its frequency spectrum (i.e. its bandwidth) into sub-channels. Time division multiplexing (TDM) assigns each low-speed device in suuccession a time slice of the bandwidth.

multiplexor (MPX) A piece of equipment used within a data communications network to concentrate the traffic from a number of transmission lines on to a single high-speed line. Multiplexors are installed one at each end of the high-speed line. They divide up the capacity of that line into a number of channels, directing the traffic from each of the slower lines into one of those channels. Thus the bandwidth of the high-speed line must be equal to the sum of the bandwidths of the slower transmission lines.
See also *statistical multiplexor*.
Compare *concentrator*.

multiplexor channel A channel within a computer system that allows many simultaneous transfers in either direction. Multiplexor channels are used to transfer data between the memory of computer systems and low-speed peripherals such as printers, and between memory and a number of low-speed communications links. Data is transferred a character at a time, but is directed to or from the correct locations in memory so as to be assembled into messages or blocks belonging to individual devices attached to the channel.
Compare *direct memory access*; *selector channel*.

multipoint The arrangement of a transmission line in which a number of terminals are attached to it, either at intermediate points along the line or on branches which radiate from it. Multipoint lines are used to reduce costs in networks of terminals using leased lines. Because sections of the line are shared between some or all of the terminals attached to the line, special protocols called *poll/select* procedures are used to control traffic flow. Compare *point-to-point*.

multiprocessor system A computer system that has more than one independent processor. There is only one area of memory, accessible to and shared between all the processors. These systems have one multiprogramming operating system, with special additional logic that distributes the work of running applications programs and the operating system itself across the different processors, to get as much work out of them as possible. This has a cost, and a two-processor system typically is capable of only 175% as much as a single-processor system, with losses increasing as the number of processors increases. However, multiprocessor systems do have the advantage of resilience, since processing can continue (at a lower level) even when a processor fails. They are used for very large single applications and where continuity of processing is important, such as for airline reservation systems or dealing systems.

Compare *parallel processing*.

multiprogramming Of a computer system that is able to run a number of programs concurrently. Multiprogramming is supervised by the operating system which shares out processor time among the programs that are running. This is worthwhile because most programs spend a great deal of their time waiting for transfers of data to and from peripherals, during which time they are not using the processor. On a multiprogramming system, programs are each assigned a priority and the processor is allocated to the program with the highest priority that is not waiting for a peripheral transfer.

multitasking Strictly speaking, refers to a software technique which enables an applications program to split itself up into a number of tasks, which can then be started up to run in parallel. Multitasking is used particularly in teleprocessing systems to permit shared access to mass storage files. Rather than wait until an applications program has finished processing one transaction before starting it up again to process the next, the teleprocessing monitor creates a new 'task' –effectively a new version of the applications program – for each transaction that arrives. While one or more of these tasks are waiting for records to be retrieved from mass storage, other tasks can continue processing. The result is more effective use both of the processor and the mass storage devices, and faster response at the terminals. Multitasking is also used in some personal computers to enable the user to continue using the keyboard while other lengthy activities such as printing take place.

Now increasingly used as synonymous with multiprogramming, meaning running several different programs at once.

multithreading A software technique used in some teleprocessing systems to enable many like transactions to share a single copy of the applications program without interfering with one another. Therefore, each new 'thread' that handles a newly arrived transaction requires only a fresh copy of the data that the transaction uses, rather than both the data and the program code as well. This makes more efficient use of memory. For multithreading to work, it is essential that the applications program should not modify itself in the course of processing so that the code can be reused safely. Such a program is described as reentrant.

musical instrument digital interface (MIDI) A protocol (or a device that implements the protocol) that enables a number of electronically-driven musical instruments to be connected to, and controlled by, a computer. It has been agreed as a standard by musical instrument manufacturers.

MYCIN The oldest and best known example of an expert system, used to diagnose infectious diseases. The doctor conducts a dialogue with the system to describe the patient's symptoms, following which the system suggests the most appropriate treatment.

N

NAK Abbreviation of *negative acknowledge*.

name service See *directory service*.

NAND circuit A circuit with two or more inputs and one output whose output is high if any one or more of the inputs is low, and low if all the inputs are high.
   See also *logic element*.

narrative See *comment*.

National Bureau of Standards (NBS) A US Federal body that has been responsible for formulating standards and for promoting their use, particularly within the public sector. Recently renamed the National Institute of Standards and Technology.

National Computing Centre (NCC) An organisation in the UK, originally government-supported and whose role was to promote the effective use of computer technology. It provides information and training services to industry and commerce.

National Institute of Standards and Technology A US Federal body that is responsible for formulating standards and for promoting their use, particularly within the public sector. Formerly called the National Bureau of Standards.

National Telecommunications Agency See *PTT*.

natural language interface (NLI) A means of controlling a computer application that relies on commands with a syntax and vocabulary similar to normal language. In theory, such a natural language interface should allow any user who knows what a database contains to find what they want and display this in the form they want, without needing to know about the internal structure of the database nor learn a set of special commands.

natural language understanding Machine understanding of normal continuous speech. This involves three levels of interpretation:
- (1) syntactic, to clarify the grammatical relationships between words in sentences;
- (2) semantic, to assign meaning to the various syntactic constituents;
- (3) pragmatic, to relate sentences to one another and to their context.

NBS Abbreviation of *National Bureau of Standards*.

NCC Abbreviation of *National Computing Centre*.

near letter quality (NLQ) Used of matrix printers that produce characters by means of multiple passes across the page, with each imprint displaced slightly to create a better letter image. Printers such as daisy wheel printers that produce a character in a similar way to a normal typewriter, by means of a preformed shape, are described as letter quality printers.

needle printer See *matrix printer*.

negative acknowledge (NAK) A control character used in binary synchronous communications to indicate that a message has not been received correctly. On receiving this character in reply to a message, the sending device will retransmit the message, continuing to do so either until it receives an acknowledge character or until it concludes, after a certain number of retries, that transmission is impossible.

nest Place one resource within another of the same type, such as, in programming, a loop within a loop or a subroutine within a subroutine.

network A collection of objects or concepts that are interconnected, either physically (e.g. communications network, local area network) or logically (e.g. associative network).

Also used in the abstract sense to mean modes of interaction between people.

network architecture A term used to mean the way in which a data communications network is put together, in other words its topology (the arrangement of switching nodes and interconnecting transmission lines) and internal structure (the interfaces and protocols that are adopted). Sometimes also used to mean the products that put the architecture into practice. The term first achieved currency with the announcement by IBM of systems network architecture (SNA) in 1974.

network attachment A term used by telecommunications authorities to refer to the procedures used to approve devices for attachment to public networks. Control of network attachment is intended to protect the network itself against damage, and protect users of the network against interference from faulty or badly designed equipment. Also known as *homologation*. In the UK, network attachment is controlled by the British Approvals Board for Telecommunications (BABT).

network control See *network management*.

network interface unit A device used to connect terminals or computers to digital transmission lines. In other words, the digital equivalent of the analog modem. Sometimes referred to as *codec*, a contraction of coder-decoder.

network layer The third protocol layer of the OSI reference model, defined by ISO. The network layer routes data through intermediate nodes and networks between the originating and receiving devices.
See also *application layer*; *data link layer*; *physical layer*; *presentation layer*; *session layer*; *transport layer*.

network level See *network layer*.

network management The tasks and equipment functions that ensure that a communications network operates reliably and as required by its users. This includes monitoring the operation of the network and detecting faults so that repairs can be made quickly and effectively; collecting statistics on traffic so that extensions to the network can be foreseen; reconfiguring the network to cope with failures and to accommodate changing requirements; and providing directories of subscribers and services. On public networks, it also includes collecting information about subscribers' use of the network as a basis for charging.

network management centre (NMC) A location which is the centre for the management of a communications network or a geographical division of a network. All information about the operation of the network is routed to the centre, which is staffed by engineers and support staff and has all the necessary monitoring and control equipment.

network model A model for the structure of a database that sees it as a network. In other words, any record in the database may be related to (and linked logically with) any other record. The network data model was influential in the early stages of the development of database techniques, particularly as embodied in the CODASYL data model, and is widely represented in current database packages. It is suitable for the programming of complex applications. The *relational model*, by contrast, is more convenient for end-users wishing to retrieve data using a variety of search criteria.
Compare *hierarchical model*.

network topology The arrangement of switching nodes and interconnecting transmission lines in a communications network.
See also *bus topology*; *ring topology*; *snowflake topology*; *star topology*.

network user address (NUA) A number identifying the location from which a user of a data communications network service, such as a public packet switching network, operates. Bills and information about the service are normally directed to that location.

network user identifier (NUI) An unique identifying number given to each subscriber to a data communications network service, such as a public packet switching network. Subscribers must enter their NUI to establish their authority (or otherwise) to use the network and to gain access to services accessible via the network, such as information retrieval services.

networking (1) Used as a synonym for telecommuting.
(2) Connection of devices to one another by means of a communications network.

neural network A method of designing a computer system that attempts to mimic some functions of the human brain. Like the brain, neural network computers are made up of large numbers of simple processing elements (called neurons) which receive incoming data in parallel, then map this into a pattern representing an acceptable solution. They are particularly good at analyzing inexact problems with many variables, such as recognising patterns in data from radars or sonars or verifying signatures on cheques.

They do not depend on programming in the conventional sense, but derive their abilities by 'learning' about the problem with which they are to deal. This learning process can either be 'supervised' or 'unsupervised'. For supervised learning the network is given pairs of problems and their corresponding solutions, then a training algorithm alters the connections between neurons to ensure that the required output is produced. For unsupervised learning solutions are not provided, and the training algorithm alters the network so that similar inputs produce similar outputs.

NLI Abbreviation of *natural language interface*.

NLQ Abbreviation of *near letter quality*.

NMC Abbreviation of *network management centre*.

NMR Abbreviation of *nuclear magnetic resonance*.

node A general term meaning a point in a communications network at which several transmission lines meet. Also used to mean the piece of equipment at which the lines terminate, and which controls and/or switches the traffic flowing along them. Telephone exchanges are nodes in a telephone network, and packet switching exchanges are nodes in a packet-switching network.

noise See *line noise*.

non-blocking switch A switching device which has the capacity to handle calls on all the lines connected to it at once, without delaying any of the calls.

non-impact printer A printer that produces an image on paper other than by a physical impact on the paper. Laser, thermal and ink jet printers are examples of non-impact printers, using various technologies to produce the printed image. Contrast with *impact printers* such as *matrix* or *daisy wheel printers*, where the print image is formed by striking an inked ribbon against the paper.

non-procedural language A programming language that does not require the programmer to define the sequence of steps required to solve the problem that is to be addressed, as is necessary with a procedural language such as BASIC or COBOL. Instead, the programmer provides certain facts or statements about the problem, and the language compiler supplies the procedural logic. Report program generators are one example of a non-procedural language, and logic programming languages such as PROLOG are another.

Compare *procedural programming*.

non-voice traffic Digital traffic on a communications network – data, text, image – as opposed to analog voice traffic, in other words ordinary telephone calls. Data traffic is sometimes used in the same sense, with the risk of confusion with its narrower meaning, referring solely to traffic between computers and data terminals.

non-volatile memory A memory device that does not lose its contents when the power is switched off. The magnetic core memories used in the past are non-volatile, but the widely-used semiconductor memory devices of today (random access memory – RAM) are volatile. Other semiconductor memory technologies, such as read-only memory (ROM), do retain their contents, but this can only be fixed during manufacture or by a special process, while the contents of RAM memory varies in the normal course of processing.

NOR circuit A circuit with two or more inputs and one output whose output is high if and only if all the inputs are low.

See also *logic element*.

normalisation A formal technique used in data analysis, which breaks down complex data structures into simple two-dimensional structures. Normalisation is associated with the relational model of database structure, and is a teasing-out process believed to reveal the true attributes and relationships of data. When the process is complete, data is said to be in third normal form.

NOT gate See *inverter*.

notebook computer The successor to the laptop computer in terms of size. It offers similar performance but is typically only the size of an A4 pad of paper and weighs 3–4 kilos. The cheapest machines have no hard disk and typically are used to send information back to head office periodically, via a modem and the telephone network. More expensive machines have most of the capabilities of a desktop computer.

Compare *laptop computer*.

noughts complement See *complement*.

NTSC The US television standard, consisting of 525 horizontal lines per frame and 30 frames per second. It is incompatible with the European standard, PAL.

NUA Abbreviation of *network user address*.

nuclear magnetic resonance (NMR) The technology underlying magnetic resonance imaging.

NUI Abbreviation of *network user identifier*.

number cruncher A computer system with great computational power, with its strength lying in carrying out lengthy or voluminous calculations, rather than transferring large amounts of data to and from mass storage.

numeric keypad A device used to enter numeric information into a computer system, such as for accounting applications. It may either be built into the keyboard alongside the typewriter keys, or consist of a separate unit. Like any type of keyboard, it works by generating character codes according to which key, or combination of keys, is pressed, and sending those codes to the processor of the terminal or computer to which it is attached.

numerical control See *computer numerical control*.

O

OA Abbreviation of *office automation*.

object A packaged element of software containing both the processing logic necessary to handle an identifiable task, and any related data. Similarly, an entity in an object-oriented database together with any processing logic associated with that entity.

For example, an object called 'invoice' might contain the logic necessary to create, collect on and cancel an invoice, and would hold data such as date raised, last date for payment, amount and so on.

See also *object orientation*.

object class A collection of objects which share a number of common characteristics.

See also *class hierarchy*; *inheritance*; *object orientation*.

object computer The type of computer on which an object program is intended to run, as opposed to the type of computer on which it is compiled.

object graphics Graphics composed of a number of objects, each defined by its characteristics and position within the image or on the screen. For example, a circle might be defined by the co-ordinates of its centre, its radius, and the thickness of the line round its circumference; a line of text by the co-ordinates of the first letter, font type, size and style, and the characters it contained. Contrast with *raster graphics*, made up of an array of picture elements (pixels), each of which is either white or black (or, in the case of grey shade or colour graphics, a shade of grey or a colour respectively).

Object graphics have a number of advantages over raster graphics. For applications such as architectural or engineering design, drawings are much easier to edit. They also occupy less storage space in memory or on mass storage. Finally, graphics can be scaled and reproduced accurately independently of the resolution of the device concerned. Thus a graphic built up on the screen of a personal computer at a resolution of about 60 pixels per inch (on which curves will have a stepped appearance) can be printed by a laser printer operating at 300 pixels per inch and will take full advantage of the higher resolution (producing, for example, smooth curves).

Object Management Group (OMG) An organisation set up in 1989 to promote the theory and practice of object-oriented technology, particularly by developing and promulgating specifications and standards. Its members are computer and software manufacturers and corporate users.

See also *object request broker*.

object orientation A way of looking at processing problems and their solution in terms of 'objects'. An object has a recognizable identity from which its function can be deduced and, in contrast with conventional sofware where program and data are separated, includes both the data and the procedures and functions that operate on it. Objects cooperate by sending messages to one another. In other words each object can be seen as a kind of mini-package that deals with one distinct aspect of the total problem.

Object orientation is exemplified by graphical user interfaces, which consist of a set of objects such as windows, drop-down menus, buttons, icons and so on, each of which can be called on to perform one particular aspect of the overall task of giving the user access to the resources of the computer system.

See also *encapsulation*; *object-oriented design*; *object-oriented programming*.

object-oriented design Analysis and design of systems according to the principles of object orientation. Central to this process is to understand and describe the object classes and class hierarchy of the area of activity for which systems are to be designed.

See also *inheritance*.

object-oriented programming Programming according to the principles of object orientation. In conventional (procedural) programming, such as in COBOL or BASIC, programs are seen as a logical sequence of functions and procedures that operate on certain kinds of data. To develop a software system or an individual program, the complete logical sequence must be conceptualized, then the program procedure and the data must be designed to carry through that logic. In an object-oriented system, by contrast, objects cooperate, sending messages to one another to accomplish their overall task. Because of this different view of the total problem, object-oriented programming is claimed to have three advantages:

(1) It is a more 'natural' way of viewing the problem, since it corresponds more closely to the way people normally look at the complex problems they have to solve – first sort out subproblem A, then subproblem B, and so on.

(2) Program code can more easily be reused, since objects can easily be moved from one application to another.

(3) Programs are easier to maintain, since the effects of changes are localized and bugs are easier to trace.

The process by which objects are created is at the heart of object-oriented programming and is known as encapsulation. Another key feature is that programming languages support relationship concepts IS-A and PART-OF. With these, objects with common characteristics can be grouped into classes that react in a similar way to the messages they receive and that inherit behaviour from their ancestors in a kind of heredity chain – see attribute inheritance and class hierarchy.

See also *generalization hierarchy*; *polymorphism*.

object program A program in the form produced by a compiler, after it has processed the source program. Normally, it will be in machine code capable of being executed by a computer system, although sometimes the operating software may need to complete the translation process before it can be run.

Compare *source program*.

object request broker A central component of a reference model for object orientation produced by the Object Management Group. It makes it possible for applications running on different computers to exchange objects, such as a data file or the contents of a spreadsheet, and at the same time maintains links between objects and the applications using them.

See also *object orientation*.

object type See *object class*.

occupancy The time a transaction occupies a teleprocessing system – the time taken from arrival of the transaction in the system to completion of processing and despatch of the final message arising as a result.

See also *response time*.

OCR Abbreviation of *optical character recognition*.

octal notation A notation of numbers to the base eight, using the decimal digits 0 to 7. Octal notation is used by some computer systems to represent the contents of computer memory and storage devices in preference to binary notation, because it is far more compact and easier to interpret. This is because each octal digit represents exactly three bits, so that codes consisting of three and six bits can easily be isolated and recognised. Octal notation is used particularly by computer systems with a 24-bit word length, which means that each word can be represented by eight octal digits.

octet See *byte*.

ODA Abbreviation of *open document architecture*.

odd parity Where a parity bit is set in a data field so that the sum of its bits is odd.

Odette A standard for the formatting of trading transactions sent via electronic data interchange, used by the motor manufacturing community in the UK.

ODIF Abbreviation of *open document interchange format*.

ODP Abbreviation of *open distributed processing*.

OEM Abbreviation of *original equipment manufacturer*.

off hook The state of a telephone connection when the handset is removed from the cradle. The same condition is produced by a modem when initiating a data call.

office automation (OA) The application of computer-based devices to administrative and secretarial tasks in the office. Originally, office automation was seen as document-related, covering such applications as word processing, electronic messaging and diary management, as compared with data processing that was data-related. But office automation equipment has been linked to data processing systems on the one hand, and has been squeezed by personal computers on the other, and this distinction has become more and more difficult to sustain.

office document architecture See *open document architecture*.

office of the future See *electronic office*.

office systems See *office automation*.

office workstation See *workstation*.

offline Of any part of a computer-based system that is not under the direct control of the main computer system involved, normally the one running the applications programs. Thus a data entry terminal would be described as offline if it recorded data locally on magnetic tape, which was subsequently taken or transmitted to a computer system for processing.
Compare *online*.

offline processing Processing carried out separately from and not under the control of the main computer system. For example, data may be written to a magnetic tape which is then transferred to a separate machine which takes the data from the magnetic tape and prints special documents. This may be done because the offline equipment cannot be connected directly to the main system, or because it is more convenient from an operational point of view to separate the offline processing in this way.
Compare *online processing*.

offline working See *offline processing*.

Oftel Contraction of *Office of Telecommunications*. The regulatory body for telecommunications in the UK.

OLTP Abbreviation of *online transaction processing*.

OMG Abbreviation of *Object Management Group*.

OMR Abbreviation of *optical mark reading*.

on hook The normal state of a telephone connection when the handset is in place on the cradle of the telephone. A modem may produce an off hook condition (the reverse of on hook) when initiating a data call, and restore it when the call is over.

online Of any part of a computer-based system that is under the direct control of the main computer system involved, normally the one running the applications programs. Thus the terminals connected to a transaction processing system via a network are described as online.
  Compare *offline*.

online database An information retrieval service that can be accessed from terminals (or personal computers) by dialling up over public networks. Online databases offer a vast range of information, ranging from economic or market statistics, through abstracts of scientific papers or newspaper items, to the complete text of articles or press releases.

online processing Processing carried out under the control of the main computer within a computer-based system.
  Compare *offline processing*.

online transaction processing (OLTP) Synonymous with *teleprocessing*.

online working See *online processing*.

op code Abbreviation of *operation code*.

open distributed processing (ODP) A term coined to describe a stage beyond open systems interconnection. Open systems interconnection, a term adopted by the International Standards Organisation to describe its work on manufacturer-independent standards, aims to permit computer systems and terminals to exchange information freely, regardless of manufacturer. Open distributed processing goes one or two steps further than that, implying also that systems can co-operate in a set of connected tasks, and that software can be freely exchanged between them, as well as information.

212

open document architecture (ODA) A standard, originated by ECMA and approved by ISO (which refers to it as *office document architecture*), that provides an abstract model of a document structure. Electronic documents conforming to the standard can be exchanged between heterogeneous computer systems, retaining features such as layouts, graphics and text fonts and styles. Optionally, they may also be transferred in a form that enables them to be processed by the receiving system. The format into which documents are converted for transfer is known as ODIF – open document interchange format.

open document interchange format (ODIF) A standard for the formatting of compound documents for exchange between dissimilar computer systems. A compound document is one containing content of different types, such as text in character form, raster graphics or object (engineering) graphics. The standard falls within open systems interconnection (OSI) standards, and in particular within open document architecture (ODA).

OPEN file A command that must be issued by an applications program before processing a data file, and particularly mass storage files. The open file command gives the operating software the information it needs to locate the required file on mass storage, if necessary asking the user (of a personal computer) or the operator (of a shared computer) to load storage volumes on which it expects to find the file. It then allocates any internal resources needed to process the file, such as memory buffers to hold incoming and outgoing blocks of records.

Open Software Foundation A group of hardware manufacturers who are cooperating to establish 'open' equipment standards to be implemented on their own and other manufacturers' equipment. The group includes the computer market leaders, IBM and Digital Equipment. The effect of open standards will be to make it easier for user organisations to use computers and software from many different vendors.
See also *open systems interconnection*.

open standards See *open systems interconnection*.

open systems Equipment (hardware and software) not limited in its application by technological barriers, particularly in terms of its ability to interwork with other equipment. In the context of its work on standards the IEEE has defined what is needed to bring this about: 'a comprehensive and consistent set of international information technology standards and functional standards profiles that specify interfaces, services and supporting formats to accomplish interoperability and portability of applications, data and people.'
See also *open systems interconnection*; *POSIX*; *Unix*.

open systems interconnection (OSI) A term adopted by the International Standards Organisation (ISO) to describe its work on manufacturer-independent standards for

213

interconnecting computer equipment, often abbreviated to OSI. At present, computer systems and terminals supplied by different manufacturers cannot easily be mixed in a single network, because they use different procedures to exchange information (known as *protocols*) and different internal formats for information that is to be processed. In defining OSI standards, the aim of ISO, and of the national standards-making bodies represented there, is to permit computer systems and terminals to exchange information freely, regardless of manufacturer. If their aim is realised, users will be able to choose freely from the equipment on offer from different suppliers, without worrying about whether items of equipment are compatible with one another.

Work on defining standards is guided by a framework, known as the OSI reference model, which defines the relationships between the hardware and software components within a communications network.

Compare with *systems network architecture* (SNA) and *DECnet*, which are proprietary interconnection schemes defined by leading computer suppliers IBM and Digital respectively.

See also *Open Software Foundation*; *X/Open*.

operand In general, something that is to be the object of a processing operation. In the BASIC statement 'A = B + C', for example, A, which is to receive the result of adding B and C, is the operand of the statement, and B and C are the operands of the addition.

More specifically, the part of a program instruction that specifies what is to be used to carry out the operation. In other words 'do a (the operation) with b (the operand) to c (the address)'. In this context, the operand is normally either a register or accumulator, or an immediate data value.

operating instructions Instructions provided as part of program documentation or with a job that is to be run, explaining how it should be operated – what storage media should be loaded; what messages should be entered via the keyboard; and so on.

operating lease A type of lease for computer equipment under which the lessor 'rents' out the equipment for an agreed period – most often three years. At the end of that period the lessor retains the equipment and (in theory at least) makes a profit by selling it on the second-hand market. The advantage of such a lease is its flexibility – it allows the lessee to upgrade the equipment under the lease or even to terminate the contract prematurely. It also has tax advantages since the lessee need not display the purchase as an asset on the balance sheet.

Compare *finance lease*.

operating software Operating software, sometimes known as system software, is the software that controls the operation of the hardware that a computer system comprises. It controls the operation of other software components running on the system and provides them with a range of services that they can call on to use the various resources available within the system, or to communicate with one another. This both simplifies their task and makes sure that those resources are used efficiently. It also provides a means for the user or a specialist operator to reconfigure after failure; to take components into or out of service; to monitor performance, and so on.

The operating software can therefore be seen in functional terms as a layer residing between the hardware on the one hand, and the applications programs, system building tools and utilities on the other. On small systems such as personal computers, the operating software consists of a single component – the operating system. On larger systems the operating software may be subdivided into further layers, with the operating system at the bottom next to the hardware and on top of that 'middleware' – supporting programs carrying out particular forms of processing, such as managing database files or controlling networks of data terminals.

operating system (OS) A special program loaded into a computer whenever it starts up. It controls access to all the resources of the computer and to its peripherals, and supervises the running of other programs. Originally, the operating system was supplied by the designer and manufacturer of the computer hardware, but more recently so-called 'plug-compatible' suppliers have begun to supply hardware alone that uses a competitor manufacturer's operating system. More recently still, operating systems have been developed independently of hardware design and manufacture, such as the CP/M and MS-DOS operating systems for personal computers, developed by software specialists Digital Research and Microsoft respectively, and the portable (i.e. useable on a number of different types of computer) operating system Unix, originated by AT&T.

Operating systems vary considerably in terms of the type of hardware they are designed to control, covering a range from the largest mainframe computers serving hundreds of users down to single-user personal computers. They also vary in terms of the services they provide for the people using them and for the programs they supervise. Of the services provided for people, it is worth mentioning three in particular:

(1) Job control language, used by programmers to instruct the operating system of large shared systems how programs are to be run.

(2) Moving down to the other end of the size scale, presentation management, a term used to describe features such as those found in Apple's Macintosh personal computer that make the applications it runs easy to learn and to use.

(3) File management, so that users can easily create and manage files containing data or programs.

operation code (op code) The code or bit pattern within a program instruction which uniquely identifies each of the processing operations included within the instruction set.

operational requirement See *functional specification.*

operational research (OR) An applied science that sets out to quantify problems in statistical or mathematical terms and thereby to find a solution, or a better solution, to the problem. Operational research scientists often work on computer systems in cooperation with systems analysts, using methods of analysis such as linear programming and simulation.

operational systems Information systems that support the day-to-day operations of an organisation – manufacturing, logistics, invoicing, financial accounting, etc. – as opposed to longer term activities such as planning, research, product development, and so on.

operator (1) A person whose job is to operate a computer system – see *computer operator*.

(2) The part of a program instruction that specifies what processing operation is to be carried out, otherwise known as the operation code. In effect, instructions say 'do a (the operator) with b (the operand) to c (the address)'.

(3) Similarly, a symbol or keyword used in program statements to link two or more operands in an expression. These operators fall into two main groups: i) arithmetic operators, such as +, -, /, etc. and ii) logical operators such as AND, OR, NOT. Operators usually (but not always, depending on the syntax) come between the operands to which they apply.

operator command An instruction to the operating system, entered by a computer operator.

operator console The device used to control most computer systems larger than personal computers. It has a screen and keyboard and is used by computer operators to enter messages, and by the operating system to reply to those messages and to inform the operators about the status of the system and of programs running under its control.

operator intervention See *operator command*.

operator precedence The sequence in which arithmetic operations within an expression are carried out. Normally:

(1) operations within parentheses are carried out before those outside parentheses;

(2) exponentiation is carried out first, followed by multiplication and division, then addition and subtraction;

(3) with operators of equal precedence, they are taken from left to right.

operators' log See *console log*.

optical bistability A property of crystals on which optical switching is based. Scientists at Bell Laboratories discovered that crystals can exist in two states, one of which lets in more light than the other. These two states can be used to represent the zeros and ones of binary arithmetic, and can be switched very rapidly using laser beams.

optical character recognition (OCR) Techniques for recognising printed or typed characters, based on their appearance (contrast with *magnetic ink character recognition*, which relies on magnetic fields). First the characters to be identified are scanned (see *image scanning*). Then the image of each character is analyzed (see matrix matching and feature extraction) and, if recognisable, is then converted into the equivalent internal character code. Characters that cannot be matched may be displayed on a screen for an operator to enter the correct character via a keyboard. Modern OCR readers are capable of reading successfully documents containing a mixture of fonts in different sizes and styles.

optical computing See *optical switching*.

optical disk A mass storage medium that holds information recorded digitally by means of lasers. Diode lasers record bits on the disk by changing the reflectance of the disk's laser-sensitive layer. Some optical storage devices can only record information on a disk once (see *WORM* – write once read many times); with others, information can be erased and rewritten as on a magnetic disk – see *erasable optical storage*. Optical disks can hold very large quantities of information – several thousand million bytes or 50,000 page images on a 12" disk.

See also *document image processing*.
Compare *videodisk*.

optical fibre See *fibre optics*.

optical filing See *document image processing*.

optical mark reading (OMR) The reading, by electronic means, of marks made by hand in pencil or pen on sheets of paper. Mark reading is used where large quantities of relatively straightforward information need to be collected away from the office environment. It has been widely used, for example, by public utilities to collect meter readings or details of appliances from households. It is also used by teaching institutions for testing purposes – students mark boxes on the test papers to indicate their answers, and the test papers can then be read and marked automatically.

Mark reading devices work in a similar way to optical character readers. The document is scanned pixel by pixel and the lightness or darkness of each pixel is measured and recorded, to build up a digital picture of its contents. Whereas an optical character reader then attempts to recognise characters by their shape, a mark reader only attempts to establish whether or not marks have been made in predefined positions on the document. Normally, the mark reading document is preprinted with boxes in which marks may be made.

optical memory card A plastic card, similar in size to a credit card, on to which information has been written by laser. The card is capable of holding several million characters of information which can be read electronically by a terminal, but which cannot be changed once it has been recorded on the card. Optical memory cards are suitable for holding large quantities of personal information that does not change very quickly, such as medical records.

optical storage Mass storage that uses optical means to record digital information, rather than magnetic means such as on conventional disk drives. Digital optical storage comes in four main forms – compact disk (CD-ROM and CD-I); WORM (Write Once Read Many times) for archival storage; erasable optical disk; and optical memory cards. All involve the use of lasers, to generate heat (for writing) and light (for reading) in a very small space. Similar technology is used for videodisks, but which hold information in analog form.

See also *optical disk*.

optical switching A means of building the logic for digital computers using laser beams rather than electronics. The principle is that laser beams are used to switch bits on and off, rather than electronic circuits. Since light can travel much faster than electrons, this would make it possible to build considerably faster computers. Optical switching could also revolutionise visual display terminals.

See also *optical bistability*.

OR Abbreviation of *operational research*.

OR circuit A circuit with two or more inputs and one output whose output is high if any one or more of the inputs are high.

See also *logic element*.

OR operation A program instruction or statement that performs a boolean OR operation on two data fields. As shown below, a one bit is set in the output field if either of the bits in the input fields is one, otherwise the bit is set to zero.

Input		Output
0	0	0
0	1	1
1	0	1
1	1	1

Oracle The trade name for ITV's teletext service.

order code See *instruction set*.

origin The absolute address in computer memory where a program or a data area starts.

In multiprogramming systems, a program is loaded into a free area of memory, and its origin is set to the start of that area. Subsequently, when program instructions generate references to memory locations within the area occupied by the program, the addressing hardware automatically adds on the origin value, to compute the absolute address of the memory location to be accessed.

original equipment manufacturer (OEM) A manufacturer that buys in computer systems in order to incorporate them into its own products, sold under its own name.

orphan A word or short line left dangling at the bottom of a page or paragraph. Used in word processing and desktop publishing applications. Compare *widow*.

OS Abbreviation of *operating system*.

OS/2 The operating system for IBM's PS/2 range of personal computers, introduced to replace the PC range. It has been developed jointly by Microsoft, originators of the MS-DOS operating system, and IBM, and is likely to become an industry standard.

OSI Abbreviation of *open systems interconnection*.

OSI reference model A conceptual framework for the development of communications networks involving computer systems and terminals, defined by the International Standards Organisation. It defines the relationships between the hardware and software

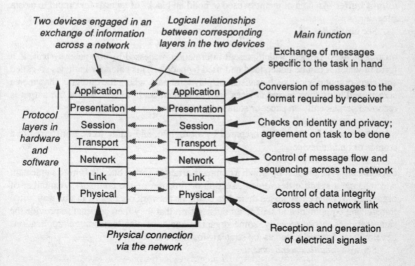

219

components within a communications network, and is intended to form the basis for standards defining protocols (in other words procedures for exchanging information) and interfaces. The idea is to bring about a situation where, unlike today, equipment from different suppliers can be mixed and interconnected freely.

The reference model defines seven layers, as shown in the figure, each of which handles particular aspects of the overall task and has a clearly specified interface with the layers above and below it.

outline flowchart See *macro flowchart*.

outline font A font that is represented within the computer as a series of lines and curves, rather than as a bit map. This means that it easier to scale the font to display it or print it in different sizes. Also known as *vector fonts*.

outliner A software package that helps a writer to develop an outline of a document prior to attempting to draft the text in full.

output (1) Transfer data or programs out of the memory of a computer system to one of its peripherals.

(2) Also used to refer to the information that is transferred in this way.

output area See *output buffer*.

output buffer An area of memory used to build up blocks of records for output to a data file.

See also *buffer*.

output devices Devices which present information processed by computer systems in a form in which it can be understood and used by people. This category includes so-called soft-copy output devices such as visual display terminals, where the output is shown on a screen, and hard-copy output devices such as printers and plotters, where the output is recorded permanently on paper or some other medium.

output record A data record prepared by a program, and ready to be transferred to a storage or output device.

outsourcing The process by which a company arranges for an outside firm to implement and manage a specific department, section or function of the company. A number of organisations have externalised their data centre or network operations in this way, often transferring equipment and staff to an outside firm that they then contract to provide the same services. More recently, some organisations have also 'outsourced' applications development, formerly seen as too strategic for such treatment.

See also *facilities management*.

overflow (1) A condition that occurs when an arithmetic calculation performed by a processor produces a result too large for the memory location or area in which it is to be stored. Programs can test for this condition and branch to a routine that takes remedial action.

(2) A condition that occurs when there is insufficient space in the home area of a direct access file to store a record, so that it is stored in an overflow area. The home area is the area where records are stored based on addresses calculated directly from some value contained within the record itself, meaning that they can normally be retrieved in one access to the storage device. When they have to be stored in the overflow area, a 'tag' or pointer is stored in the home area, which means that more than one access is needed to retrieve them.

overflow area The area of a direct access file where records are stored for which there is not enough space in the home area.

overlay A section of a program that is called into memory (usually from disk or some other kind of direct access storage) when it is needed, rather than being resident in memory all the time a program is running. Overlays are used to reduce the minimum amount of memory needed to run a program, and to economise on the amount of memory occupied by a program in a multiprogramming system. The main penalty is increased running time, because extra time is needed to load overlays into memory from storage when they are needed.

own coding Small sections of program code that the user can incorporate into a standard program such as a utility or an applications package, to deal with special conditions such as non-standard file or record formats.

P

PABX Abbreviation of *private automatic branch exchange*.

package An off-the-shelf computer program.
See also *application package*.

packaged software See *software package*.

packed decimal See *binary coded decimal*.

packet The smallest unit of information transmitted by a packet-switching network. The packet has a standard format and a maximum length set for a particular network. This is designed to enable the network to handle traffic as efficiently as possible. Messages submitted by computer systems or terminals for transmission over a packet network may be longer than the permitted maximum packet length, in which case they are split up into packets by the network software and sent separately. The network software also re-assembles them into a complete message on arrival at their destination.

packet assembler disassembler (PAD) A feature built into a packet switching network, usually referred to as a PAD, that allows devices that are not capable of full packet switching protocols to use the network. PADs are special ports in the nodes of packet-switching networks that can be dialled up over the public telephone network. They were originally included for the use of teletype-compatible keyboard terminals, but are now used by a wide range of devices, including personal computers. They convert messages received from these devices into packet format and vice versa.

packet mode terminal A terminal (in the telecommunications sense, that is any device that can be attached to a data communications network, including computer systems) able to execute packet-switching protocols. Packet mode terminals can be attached directly to the nodes of packet-switching networks; can answer and initiate calls; use all the facilities of the network; and send multiple streams of data across the network at high transmission speeds. *Character mode terminals*, by contrast, must be connected via a device in the node called a packet assembler/disassembler (PAD); can initiate but not answer calls; and can only use a subset of the network's facilities at relatively low speeds.

packet radio Packet switching using portable radio transmitter/receivers, rather than conventional data terminals, to exchange messages over the network. Carrier sense multiple access (CSMA) protocols are used to ensure that messages broadcast by the packet radio devices do not collide. Packet radio has advantages where users are highly mobile, such as in a military operation.

See also *mobile data communications*.
Compare *cellular radio*.

packet switching A method of handling messages transmitted across a data network. It uses a form of switching in which the two devices engaged in the exchange of messages share the transmission path between them with other devices using the network. As far as the two devices are aware, they are connected by a transmission path of which they have full and exclusive use until the call terminates. This is known as a *virtual circuit*. Messages sent across the virtual circuit are split up into packets of a convenient size for

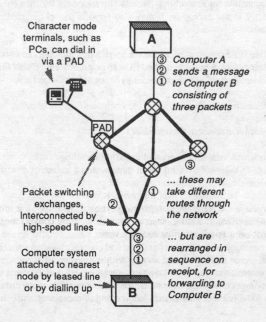

Character mode terminals, such as PCs, can dial in via a PAD

Packet switching exchanges, interconnected by high-speed lines

Computer system attached to nearest node by leased line or by dialling up

③ ② ① Computer A sends a message to Computer B consisting of three packets

③
... these may
① take different routes through the network

③
② ... but are
① rearranged in sequence on receipt, for forwarding to Computer B

223

transmission – typically several hundred bytes – and are reassembled on arrival.

A packet-switching network consists of a number of packet-switching exchanges interconnected by high-speed transmission lines. The exchanges are responsible for: routing the packets across the network, choosing the best route based on the traffic on the network at the time; storing packets temporarily until a transmission path is available; and ensuring that packets arrive at their destination uncorrupted and in the same sequence as they were sent. They may also contain PADs (packet assembler disassemblers) – special ports which enable simple character mode terminals such as personal computers to use the network.

Packet switching makes efficient use of transmission capacity where a large number of computer systems and terminals wish to share transmission paths between one another. Many countries have now installed national public services. These are interconnected so that data calls can be made across them in much the same manner as international telephone calls.

See also *X.25*.

Compare *circuit switching*; *message switching*.

packet-switching exchange (PSE) A switching and connection point in a packet-switching network. Each exchange in a network routes packets to and from other exchanges, and is also responsible for assembling packets into messages for forwarding to devices connected directly to the exchange. It may also provide packet assembler disassembler (PAD) facilities.

packet-switching service (PSS) A data transmission service that uses packet-switching techniques. British Telecom's own service in the UK was originally called this but has been renamed Packet SwitchStream, which continues to fit the abbreviation PSS.

packing density See *recording density*.

PAD Abbreviation of *packet assembler disassembler*.

page (1) The areas into which virtual memory is divided – see *paging*.

(2) A unit of storage in videotex systems, or in other words a screenful of information. Also known as a *frame*.

page description language A programming language used to instruct an intelligent printer (such as some laser printers) what to print, or in other words software for graphically composing images on a page. Until recently, printers were usually driven by a stream of characters, consisting of printable characters and control characters specifying paper movement, boldface or underline, and so on. With the arrival of more versatile printers such as laser printers, that can print graphics as well as text, and text in a whole range of typestyles, languages have been defined which make it possible to describe with great precision what is to be printed. Rather than just a string of characters, the printer is sent a special program that it interprets in order to determine what to print. The program can instruct it, for example, to produce a circle of a specified diameter in a particular position on the page; to skew or rotate an image a specified number of degrees; to fill a specified area with a shade of grey.

Although people can and do write programs in these languages, which can then be sent to the printer to be executed, normally they are used internally by applications packages such as word processing or drawing programs, to convert the document or drawing that the user has built up on the screen into commands that can be interpreted by the printer. This means that higher print quality can be achieved on the page than can be displayed on the screen.

The most widely used page description language is Postscript, a general-purpose language originated by Adobe Systems, a company specialising in typefaces for computer-generated documents. It is supported both by desktop laser printers and by typesetting machines. A number of other languages are in use, including for specialised applications such as computer-aided design.

page printer A printer that forms and prints a page at a time. The most important device in this category is the laser printer, which uses lasers to build a page image, dot by dot, then an electrostatic printing engine, such as in a photocopier, to transfer the image on to paper.

Compare *character printer*; *line printer*.

paged memory management unit (PMMU) A component of advanced microprocessor chips, that helps the operating system to manage computer memory.

See also *paging*; *virtual memory*.

pages per minute (ppm) A measure of the speed of page printers.

paging A technique used in computer systems with virtual memory. Large areas of program or data are divided up into pages, usually of a fixed length. When memory occupied by the program or by the data is required for use by other programs, pages are written to disk storage temporarily, and are read back into memory again when needed once more. This technique makes efficient use of computer memory, space in which is more expensive than space on disk.

PAL Abbreviation of phase alternate line, a standard for colour television broadcasting adopted in the UK, West Germany and elsewhere. It consists of 625 horizontal picture lines per frame and 25 frames per second. Compare with *SECAM*, another TV standard used in Europe, and *NTSC*, the US standard.

palette See *colour gamut*.

palmtop computer A personal computer small enough to fit inside a jacket pocket and typically weighing 500g or so. They usually come with special-purpose applications software but can also run cut-down versions of standard personal computer packages. Their miniature keyboards and small screens make them suitable for limited information retrieval (such as from a diary) and for note-taking.

Palo Alto Research Centre (PARC) A research and development unit established by Xerox Corporation. It is noted in the information technology field for two important technological innovations, both introduced in the 1970s.

(1) Its work on improving the interface between people and computers gave rise to the first graphical user interface, embodied first of all in the Xerox Star workstation and later, with greater commercial success, in Apple's Macintosh personal computer.

(2) It developed the first local area network, Ethernet, subsequently licensing the technology widely to other manufacturers.

pantone A standard set of colours identified by numbers, used within the printing industry.

paper control loop A loop of paper contained in line printers that controls the movement of the continuous stationery to head of form and to preset positions within each page. The loop is normally the same length as one page, and has holes punched in different channel positions across the loop at various points along its length. Applications programs can send commands to the printer to advance the paper to the hole in a specified channel in the loop. By convention, channel 1 is used to mark the top of the page.

paper tape punch An output device, now obsolete in computer systems, that encodes data into paper tape by punching holes in the tape. Different combinations of holes in channels (usually eight, consisting of seven data bits and one parity bit) across the tape were used to represent alphanumeric characters.

paper tape reader An input device, now obsolete in computer systems, that reads punched paper tape – see *paper tape punch*.

paper throw The movement of paper through a printer other than by single line spacing. Paper throw is used, for example, to move the paper to the top of the next page after the last line on the preceding page has been printed.

paperless office See *electronic office*.

parallel computing See *parallel processing*.

parallel importer See *grey importer*.

parallel interface A method of transferring data between a peripheral and the memory of a computer system in which all the bits that make up a character of data are transferred simultaneously. Also used to refer to the device which executes the transfer.

Compare *serial interface*.

parallel processing Electronic processing of information by carrying out a number of operations in parallel, rather than sequentially as in a conventional computer (see von Neumann architecture). Array processors, used for applications such as weather forecasting, are one example of parallel processing. Parallel processing is also seen as a key technology for knowledge-based systems, which often involve time-consuming passes through large volumes of information. Two main architectures are being researched – single instruction multiple data and multiple instruction multiple data.

parallel run Where a new computer-based system is run in parallel with the procedures it is intended to replace. The results produced by the new system are compared with those produced by the old in order to verify that it is working correctly.

parallel transmission See *parallel interface*.

parameter A variable passed between a procedure or program, and the routine – a function or a subroutine – that it is calling. Sometimes referred to as an argument.

PARC Abbreviation of *Palo Alto Research Centre*.

parity bit An extra bit added to a data field to detect when the field has been corrupted, such as in transmission. When ASCII characters are transmitted, for example, each character occupies seven bits and the eighth bit is used as a parity bit. The value of the bit is set according to the sum of the bits in the field. If the sum, including the parity bit, is even, it is known as even parity. Similarly, where the sum is odd it is known as odd parity.

parity check A check applied to a character or a series of characters, based on the addition of redundant bits called parity bits. Parity checks are applied to detect corruption of data.

partition (1) A fixed area of memory, set aside in a multiprogramming system for the use of a program. Early multiprogramming systems used to allocate their memory in this way, but modern systems allocate memory dynamically.

(2) A subdivision of a file or database, normally determined on a logical basis such as to hold a specified range of key values. Files or databases may be partitioned so that parts of the files can be held at different locations (see distributed data processing) or to minimise the effect of failures – when a failure occurs only the partition that is affected need be recovered (see file recovery).

Pascal A high-level programming language designed by Nicklaus Wirth, and introduced in the late 1960s. Its main ancestor is Algol and, like Algol, it is a block-structured language. The advantage Pascal has over Algol is that it can easily be compiled on small computer systems and produces very efficient programs. There are several different versions of Pascal. Apart from the version defined originally by Wirth, a modified version developed at the University of California in San Diego (known as UCSD Pascal) is widely available on personal computers.

Sample Pascal program

```
Program GuessAgain ;
-- Asks user to enter first a target number, then the
-- largest number with a square less than that.
-- Rejects incorrect guesses; stops when user replies 'N'.
Var Answer : Char ;
Procedure PerfectSquare ;
Var Correct, Guess, Target : Integer ;
Begin
  Write ('Type in a target number: ') ;
  Readln (Target) ;
  Correct := Trunc (Sqrt (Target) ) ;
  Write ('Now guess the largest integer whose square
       is no larger than ' , Target , ':' ) ;
  Readln (Guess) ;
  While Guess <> Correct Do
  Begin
    Writeln ('No – that gives ' , Guess * Guess) ;
    Write ('So guess again: ') ;
    Readln (Guess)
  End ;
  Writeln ('Good ' , Guess , ' has the largest square
       no bigger than ' , Target , '.' )
End; (* Perfectsquare *)
Begin
  Repeat
    Perfectsquare ;
    Write ('Do you want to try another target? ') ;
    Readln (Answer)
  Until Answer = 'N'
End.
```

---

passive Used of systems that do not take the initiative, but rely on the discretion of their human users. A passive electronic mail system, for example, is one that puts a message in an electronic mailbox and waits for the subscriber to collect, rather than attempting to forward it to the subscriber immediately.

Compare *active*.

passive matrix See *supertwist*.

passive star A star topology for a network in which the branches are connected physically at the central point. Signals introduced at any point on the star therefore travel through all the wire in the network.

Compare *active star*.

password protection The use of passwords to prevent unauthorised access to computer systems. When they attempt to gain access to a computer system from a data terminal, users are required both to identify themselves, such as by entering a user number or name, and also to enter a password that is associated with the user identifier. If the password is correct, they are allowed access to the system, possibly restricted only to certain services according to their authorisation level. Various additional mechanisms may be used to increase security still further, such as by requiring users to change passwords periodically or by using more than one level of password.

patch Make an amendment directly to an object program or to a data file to correct an error, to avoid going through the normal amendment process, such as changing the source program, recompiling and so on.

patch panel A device used to change the connections between pieces of equipment, by fitting plugs into sockets on a panel. Patch panels are used, for example, to change the connections between modems and the ports in a computer, so that faulty network components can be bypassed.

pattern recognition Recognition by electronic means of patterns in data. Pattern recognition is applied to images received from radar or sonar, and also to images that have been processed by a scanner and expressed as a bit map. See for example optical character recognition.

PBX See *private automatic branch exchange*.

PC Abbreviation of *personal computer*.

PC-DOS The operating system for the IBM PC and compatibles. See *MS-DOS*.

PCB Abbreviation of *printed circuit board*.

PCL Abbreviation of *printer command language*.

PCM Abbreviation of *plug-compatible manufacturer*.

PCM Abbreviation of *pulse code modulation*.

PCMCIA Standing for *Personal Computer Memory Card International Association*, acronym is used to describe an expansion slot provided in notebook and handheld computers to accept memory cards. The original version of the standard has since been extended to support miniature hard disks as well, raising the possibility that desktop computers will incorporate a similar slot so that the hard disk can easily be transferred from notebook to desktop and vice versa.

PCTE Abbreviation of *portable common tools environment.*

PD Abbreviation of *public domain.*

PDH Abbreviation of *plesiochronous digital hierarchy.*

PDI Abbreviation of *picture description instruction.*

PDN Abbreviation of *public data network.*

peak-to-average ratio The ratio of the normal, instantaneous speed of transmission of a particular form of communications traffic, to the average speed of transmission in a given time period. This is a measure of its 'burstiness'. Telephone traffic, which occupies the whole of a communications link for the duration of a call, has a peak-to-average ratio of one. Traffic from a personal computer attached to a local area network, by contrast, may have a peak-to- average ratio in the hundreds – in other words, occasional high-speed bursts of traffic. Different network and switching technologies perform well with traffic whose peak-to- average ratio falls within a certain range. Circuit-switching, used for telephone networks, performs well with low peak-to-average ratios, whereas packet-switching and local area network technologies are suited to high ratios, in other words very bursty traffic.

PEEK statement A BASIC statement used to obtain a value from a specified memory location.

peer-to-peer A relationship between communicating devices in which any device in a network of computers and terminals may initiate or respond to messages. Peer-to-peer protocols are beginning to replace the master-slave protocols normally used in computer and terminal networks for transaction processing, such as IBM's binary synchronous communications (BSC).
    See also *bit-synchronous*; *client-server.*

pen computer A personal computer, usually portable, that is controlled by means of an electronic pen rather than via a keyboard and/or a mouse. The pen is used to point at, write or draw on the screen or on a writing tablet, which automatically senses its position. Pen computers are used mainly for data capture in the field, such as by field engineers.
    The most innovative approach to the concept comes from a Californian company

230

called GO Corporation. GO has developed a special-purpose operating system for pen computers called PenPoint, which it has licensed to a number of other manufacturers who are developing their own pen-based products. PenPoint incorporates handwriting recognition techology and recognises a set of standard text-editing symbols, such as a caret for 'insert here'.

PenPoint See *pen computer*.

PERFORM statement A statement used in high-level languages, including COBOL, to execute a procedure or subroutine.

peripheral A component of a computer system that operates under the direct control of the processor, attached to it via its input/output logic. A peripheral can be contrasted with a terminal, which is is attached via a communications link. The term is normally used to refer to mass storage devices such as disks and tapes, and input/output devices such as scanners and printers, in other words those devices to and from which programs may transfer data. It is not normally used to refer to devices used by people to interact with or control a system, such as the screen and keyboard of a personal computer or the operator console of a larger system, which also operate under the control of the operating system.

peripheral limited See *input/output limited*.

permanent virtual circuit (PVC) A continuous connection across a packet-switching network. Normally, devices using a packet-switching network establish a connection by making a call to another device, in a similar way to telephone users. The resulting connection is known as a virtual circuit. Subscribers can arrange with the operator of the packet switching network to have a virtual circuit permanently established between two devices, such as between two computer systems that continually exchange information, thus obviating the need to make calls to establish the connection.

personal computer (PC) A computer designed for individual use, rather than intended to be shared by a number of users concurrently. The term achieved currency when IBM introduced its own best-selling product called by this name, and often abbreviated to PC. Microcomputer was the original term for personal computer and is still used in this sense, although microcomputers appear in more forms than just personal computers.

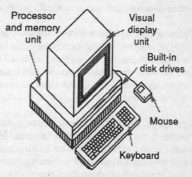

231

phase change A technology for colour printing. Blocks of coloured wax in the three primary colours plus black (see CMYK) are melted in ink reservoirs. The melted ink is then sprayed on to the paper through tiny perforations in printhead nozzles, much like ink jet printers, with all four colours applied in a single pass. Unlike the water-based inks used in typical ink jet printers, wax-based inks solidify on the page very quickly, inhibiting absorption that can reduce the sharpness of images. High-pressure rollers then flatten the dots and fuse them to the page.

Compare *dye sublimation.*

PHIGS Standing for programmers hierarchical interactive graphics system, PHIGS is a recognised international standard for applications that combine modelling techniques with computer graphics. Routines implementing the standard can be obtained as a package and called from programming languages such as C or Fortran. An extended version – PHIGS Plus – provides more sophisticated features such as surface shading and depth cueing.

Compare *graphics kernel system*; *initial graphics exchange specification.*

phoneme A sound made in human speech. Some devices which create human speech synthetically put together words and phrases by drawing from a library of phonemes stored in memory. This is known as phoneme synthesis.

Compare *formant synthesis.*

photo CD A system developed jointly by Kodak and Philips, intended as a standard means of storing high-quality photographic images on a compact disk. The aim is that normal High Street photo developers will be able to create photo CD disks from normal 35mm film by means of a CD-WO (write once) device. The photo CD format is compatible with both CD-I and CD-ROM XA formats. This means that it will be possible to view images at home on a CD-I player or a TV. It will also be possible to import them into a personal computer equipped with a CD-ROM drive in order to view or edit them or to incorporate them into multimedia publications or printed documents.

photodiode array A technology used in image scanning. Small photodiodes are arranged in a row, one for each pixel in the document to be scanned. Light is flooded on to a narrow line across the page and reflected light is collected by a lens system and focused on the diode array. Each diode produces a signal that is proportional to the light intensity of the spot focused on it, which is collected and stored. The document is scanned line by line in this way.

physical Used to describe a resource – a channel, a file, a peripheral unit, a memory location – which is identified in terms of its physical attributes, such as a hardware address or a name recorded on it. Programs often refer to resources in logical terms so that they can be run under different circumstances – on a different computer configuration, using different peripheral units, in a different area of memory – without needing to be amended. The operating software, perhaps assisted by the computer operators, translates logical references into physical terms when the program is run.

Compare *logical*.

physical layer The first and lowest protocol layer of the OSI reference model, defined by ISO. The physical layer encodes a bit stream to physically move data across the various transmission lines that form the network.

See also *application layer*; *data link layer*; *network layer*; *presentation layer*; *session layer*; *transport layer*.

physical level See *physical layer*.

physical record The form in which data is actually recorded on a storage device, as opposed to the form in which it is normally processed by an applications program – the logical record. Physical records usually contain a number of logical records, so that space on the storage device is used efficiently. Physical records on tape are also called blocks, and on disk, sectors.

physical schema See *internal schema*.

pica A unit of typographical measure traditionally used within the typesetting industry. There are approximately six picas to the inch.

Pick An operating system for small computers designed by Richard Pick. It is a multi-user operating system with virtual memory, designed around its own relational database and a high-level query language. It is regarded as particularly easy to use and versatile. In the search for a portable operating system for small computers, Pick competes with Unix, and has a smaller but equally enthusiastic group of supporters.

picture description instruction (PDI) A code used to instruct a receiving terminal to construct a graphic shape. Used in videotex systems (such as Telidon in Canada) employing alphageometric display methods.

picture element See *pixel*.

pilot system A computer-based system implemented on an experimental basis, usually to determine precisely what effects it will have before proceeding with implementation on a wider scale.

pin feed Paper fed through a printer by pins on a roller engaging in sprocket holes punched in the margins of the paper, rather than by pressure of rollers against the paper (friction feed).

233

pinouts The precise connections of the wires to the various pins in an interface socket or in a cable connecting a peripheral to a computer.

pipelining A technique used to increase the effective speed at which program instructions are executed by a processor. Instructions are normally executed by a processor serially, one instruction after another. In pipelining, the instruction is broken down into its various stages, each of which is carried out by a separate component in the processor. Each such component executes one stage of a different instruction simultaneously, which means that a number of instructions effectively overlap. An advanced processor chip powering a personal computer, for example, divides each instruction into six stages: (1) instruction fetch, (2) instruction decode, (3) address calculation, (4) operand fetch, (5) instruction execution and (6) store result.

piracy See *software piracy*.

pitch The number of characters per inch printed by a character printer such as a daisy-wheel printer. These are normally either 10 pitch or 12 pitch.

pixel Abbreviation for picture element, originally meaning the elements that are used to create a television picture and, by extension, the smallest discrete element of a bit-mapped visual display or a scanned image that can be distinguished by a computer. Both are made up of a grid of dots, each of which is either black (or a colour or shade) or white. Dots per inch is therefore synonymous with pixels per inch.

pixels per inch See *dots per inch*.

PL/1 A high-level language defined by IBM in the mid-1960s. The aim was to combine the good features of commercial languages such as COBOL and scientific languages such as ALGOL. Like ALGOL, it is a block-structured language with good mathematical capabilities; like COBOL, it has good file handling and input/output features.

plasma display A technology for visual display screens consisting of a series of gas-filled cells, one for each pixel of the display. These cells glow when a current is applied. Plasma displays are flatter than cathode ray tubes (the most widespread display technology) but have a lower resolution.

platen A solid backing against which the paper is held within a printer, usually consisting of a cylindrical roller. It is used in impact printers to form a base for the mechanism which strikes the ribbon against the paper.

plesiochronous digital hierarchy (PDH) A technology for multiplexing broadband digital signals such as those carrying video or speech. 'Plesiochronous' means 'almost synchronous' and it works by adding bits here and there to incoming signals to equalise their bit rates so that they can be combined on to a higher speed outgoing circuit. The drawback of this technology is that demultiplexing to unscramble the various signals is complex and expensive, and that operating and maintenance overheads are high.

Compare *synchronous digital hierarchy*.

plotter A device which draws on paper or film with pens, under computer control. Plotters have been used for some time to produce maps, plans and engineering drawings, and are now also being used in offices to produce overhead slides for presentations and graphs and diagrams to be included in reports.

plug-compatible Used to describe a product that is a functionally identical replacement for that of a competitor, in the sense that it can be plugged into the same interface sockets and appears to the other components of the system to operate in an identical way. However, it may differ in terms of the technology it uses internally and hence have a superior performance or a lower cost than the product it is intended to replace.

plug-compatible manufacturer (PCM) Plug-compatible manufacturers first appeared on the scene in the late 1960s and early 1970s. Unable to compete with the market leader, IBM, in offering complete computer systems, they specialised in particular high-cost components, first central processors and memory, then disk and tape drives. These were designed to work with the other hardware components and with the operating software supplied by IBM, or indeed by other PCMs. PCM competition later spread to other suppliers and to other components, such as communications processors and terminals, to the extent that a number of organisations ended up running 'IBM systems' of which few if any of the hardware components had been purchased from IBM.

plugboard A board containing a number of sockets, into which plugs interconnected with wires can be inserted to set up circuit connections. Some early computers were programmed in this way.

Compare *patch panel*.

PMMU Abbreviation of *paged memory management unit*.

point A unit of typographical measure, used in word processing and desktop publishing applications to specify the size of fonts (e.g. 10 point Times) and distances between lines and characters. There are 72 points to an inch.

point-and-click interface See *graphical user interface*.

point of sale (POS) See *electronic point of sale*.

point of sale terminal A general term for terminals used by sales assistants to capture the details of sales to customers, on the spot. Those details are usually transferred automatically to a computer system elsewhere. A range of different devices fall into this category, from electronic cash registers, through scanners used at supermarket checkouts, to transaction telephones used for credit card purchases.

point size A font's height measured in points. Generally body text fonts are 10 to 12 points. Headings are usually 14 to 30 points or larger.

point-to-point Of a transmission line that connects two devices or two locations directly.
Compare *multipoint*.

pointer (1) A small symbol on a display screen, such as an arrow or a cross, that moves in response to the movement of a mouse or tracker ball.
(2) A field in a data record that records the whereabouts of a related record. In a bill of materials file for example, a record of a sub-assembly may have a pointer (that is, a disk address) to its first part, which in turn has a pointer to the second, and so on. Files organised in this way can be accessed and updated rapidly, but the software can have problems keeping pointers up-to-date and it can become increasingly difficult to find space to add pointers as files grow.
(3) A variable used by a program to record the address of a data field whose location varies.

POKE statement A BASIC statement used to store a value directly into a specified memory location.

poll/select Procedures used to control the flow of traffic on a multipoint (sometimes known as multidrop) transmission line, in other words a line shared by a number of devices – normally data terminals. Polling is the method used to find out whether a terminal is ready to send or receive a message, and selecting is the method used to inform a terminal that the line is available for a transmission. Versions of poll/select procedures include hub polling and roll call polling.

polling See *poll/select*.

polymorphism A concept associated with object orientation, by which an object belonging to a particular class may if it wishes override the normal behaviour of its 'parent'. Thus the same message may produce different results depending on which class of object responds to it.
See also *class hierarchy; inheritance*.

pop Remove an item from the top of a stack.
Compare *push*.

pop-up menu A menu that extends upwards on the display screen when activated.
See also *pull-down menu*.

populate Establish the initial content of a database or a data file, such as details of
customers or products, so that applications programs can start using it.

port (1) As a noun, the point of access to a computer system or to a node of a communications network, such as a packet switching exchange. Normally, a port will operate at a
particular transmission speed. Devices may either be permanently connected to a port or
may call in to make a connection, and the communications link they use will operate at
the speed of the port.
  (2) As a verb, to adapt software written to run on one type of machine so that it will
run on another. Thus, a database management package written for the IBM PC might be
'ported' to run on the Apple Macintosh. Also known as *migrate*.
  See also *portable*.

port control The policing and management of access to the ports of a computer system
or network. The purposes of port control include preventing unauthorised access and
ensuring that the available ports are used as efficiently as possible.

portable (1) Of software, able to run on another make or model of computer. Used of
programming languages (such as Pascal) and operating systems (such as Unix) that are
designed to be independent of any particular type of computer system. This kind of
portability is achieved by including the software within a shell that can easily be modified to accommodate the peculiarities of the computer on which it is to run.
  Also used of applications programs that, once programmed, can easily be moved
from one type of computer to another. This may either be *source portability*, meaning
that the source program must be recompiled on the target computer, or *binary portability*,
meaning that recompilation is unnecessary.
  (2) Of personal computers or terminals, able to be carried about and operated away
from the office.
  Compare *compatibility*; *interoperability*; *luggable*.
  More detail *byte ordering*; *data alignment*.

portable common tools environment (PCTE) An integrated program support environment designed by a consortium of European manufacturers working within the
ESPRIT programme. The idea is to free software developers from working with proprietary operating systems. It is likely to be adopted as a European standard.

portrait format Used of pages printed by a word processing application where the
length of text (or other information) printed on the page is greater than its width, in other
words the normal format for a letter.
  Compare *landscape format*.

POS Abbreviation of *point of sale* .

POSIX POSIX is a standard laid down by the IEEE (coded P1003) that defines the external characteristics and facilities of a portable operating system based on Unix. It has been incorporated in a number of specifications of open standards, including the application portability profile (APP) and the X/Open consortium's common applications environment (CAE).

See also *open systems*.

post-implementation audit An audit carried out after a computer system has been in use for some time, intended to find out how closely it reflects its original specification and how closely it is meeting its original cost-benefit objectives.

post-industrial society See *information society*.

Post Office work unit A mix of instructions designed to provide a measure of the average instruction time of a processor with a typical workload, developed by the British Post Office during the 1960s. In other words, a more accurate measure of instruction time than a raw mips (millions of instructions per second) figure.

posterise A technique applied to digital images such as photographs in desktop publishing. The full range of grey scale values contained within the image is reduced to a limited set of preselected values.

PostScript A page description language originated by Adobe Systems, a company specialising in typefaces for computer-generated documents. Normally it is used by software packages, such as for graphics or desktop publishing, rather than directly by a programmer. These packages generate a PostScript program describing what is to be printed and send this to the printing device, such as a laser printer or a typesetter, where it is interpreted to draw the required image on the page. PostScript programs in turn refer to descriptions of different fonts (known as *outline fonts*), held within or sent to the printing device, which can be scaled to whatever font size is required.

Sample PostScript program

```
% program to draw a series of ellipses
/doACircle
 0 0 54 0 360 arc stroke  def
/doAnEllipse
 1 .75 scale doACircle stroke  def
300 500 translate doACircle
4  0 -72 translate doAnEllipse  repeat
showpage
```

Another version of PostScript, known as *Display PostScript*, is designed to display images on a screen rather than to generate printed output.

Compare *trueType*.

power user Someone who places heavy demands on personal computer hardware and/or software.

ppm Abbreviation of *pages per minute*.

preamble sequence A series of bits transmitted by a device to a transmission channel such as a local area network, to establish synchronisation with other devices using it.

precedence See *operator precedence*.

predicate A statement or formula that expresses a property that some object or idea might possess. In conventional computer applications, predicates usually express numeric or value relationships, such as:

$$= \text{(is)}$$
$$\text{NOT} = \text{(is not)}$$
$$< \text{(less than)}$$
$$> \text{(greater than)}$$

In knowledge-based systems, predicates may also express non-numeric relationships. In the phrase 'John likes fishing', for example, the predicate is 'likes fishing'.

See also *predicate calculus*; *logic programming*.

predicate calculus A specialised system of logical constructs capable of representing the form of arguments, so that it is possible to check in a formal way whether or not they are valid. Predicate calculus builds propositions out of terms (representing objects such as individuals, things or concepts) and predicate symbols (expressing the relationships between them). A technique called theorem proving can then be used to examine the propositions by applying rules of inference to them. As a result it is possible to determine automatically whether anything further follows from them. To take a simple example, if we have two propositions: 'Mary likes flowers' and 'John likes anything that Mary likes', we can infer also that 'John likes flowers'.

Predicate calculus can be used to express expert knowledge so that it can be incorporated in the knowledge base of an expert system. Predicate calculus also forms the basis of logic programming languages such as PROLOG, which are used to develop applications in the area of artificial intelligence.

presentation layer The sixth protocol layer of the OSI reference model, defined by ISO. The role of the presentation layer is to make sure that data is in the right format for the device that is to receive it, such as by carrying out translation between standardised data formats.

See also *application layer*; *data link layer*; *network layer*; *physical layer*; *session layer*; *transport layer*.

presentation level See *presentation layer*.

presentation management Routines included within the operating system of a personal computer that manage the exchange of information between the human user and the computer. The routines are used by applications programs and packages to do such things as manipulate windows on the screen and handle pull-down menus. It presents these and other features of the system to the user in a consistent way, and this makes the system easier to learn and to use.
See also *WIMPs*.

Prestel The trade name for British Telecom's videotex service.

preventive maintenance Maintenance of computer equipment undertaken on a regular, scheduled basis, with a view to anticipating and preventing hardware failures. Contrast with *corrective maintenance*, which is carried out after equipment fails.

price degradation Where the price that a particular product commands declines over time. This has been a feature of the market for most information technology hardware for some years. It reflects the underlying rapid improvement in the cost/performance of the key technology on which information technology hardware is based – semiconductors. Since the mid-1950s, semiconductor technology has followed a regular seven-year cycle of major innovations, each of which has resulted in major – usually orders of magnitude – improvements in the processing power available from a single component or chip.

primary memory See *memory*.

primitive A small routine held within an operating system that programs call to perform certain operations, such as to communicate with one another.

print format How information is to be laid out in a printed report. A print format shows the information (such as page numbers) to be included in page headers and footers, and the position of fixed and variable (in other words, depending on the originating program or on data file contents) fields within each line.

print position The relative position of a character within a line of a computer printout. Positions are numbered from 1 upwards, starting at the lefthand edge of the page. Line printers normally print either 120 or 132 characters on a line, using a fixed spacing between characters.

print server See *server*.

print spooler See *spooler*.

printable character A character that can be printed out on paper or displayed on a screen. Contrast with *control characters*, that are used to control the operation of devices such as printers or visual display terminals, or of transmission lines. These are also referred to as invisibles.

printed circuit An electronic circuit etched into the surface of a board made of insulating material. The process can be automated so that many copies of the printed circuit can be manufactured at low cost.

printed circuit board (PCB) The board on to which a printed circuit is etched, including the holes and mountings to take the various components that complete the circuit, such as integrated circuits, coils and so on. The term is used to mean both the board itself and the components mounted on it. Also used as synonymous with expansion card, as in 'video board'.

printer A computer peripheral that prints text, and sometimes graphics, on paper. Printers vary widely both in printing speed and in print quality. The main types of printer are:
    (1) line printers, used for high-volume printing on large computer systems;
    (2) matrix printers, used on personal computers, cash tills and data terminals;
    (3) letter-quality printers, used for word processing;
    (4) laser printers, used for word processing and desktop publishing.

printer command language (PCL) A page description language, defined by computer manufacturer Hewlett-Packard, that is supported by a range of printing and typesetting equipment.
    Compare *PostScript*.

printout A general term for the printed output from a computer system, most often used to refer to reports and lists produced by line printers on continuous stationery.

privacy The restriction of access to a computer system, or to the information it holds, solely to those authorised to have it. Also the right of individuals to control the use of personal information held on computers.
    See also *data protection*.

private automatic branch exchange (PABX) A telephone exchange installed on private premises, such as in a business office or hotel, that switches calls automatically between extensions in those premises, and also to and from the public telephone network. Also known as *private branch exchange* or PBX.

private line See *leased line.*

private network A communications network installed and (usually) operated by an organisation for the use of its own staff and computer systems. Where the private network runs between sites, it will normally use services provided by a common carrier, either leased for exclusive use or on demand. A private network may also be connected to public networks, so that calls can be made from devices attached to the private network to the outside world, and vice versa. To protect the monopoly of the common carrier, there are normally strict rules (varying from country to country) as to the types of traffic which may and may not be carried over a private network. As a general rule, third-party traffic, in other words traffic which neither originates nor terminates on the private network, may not be carried over a private network except by organisations holding a special licence.

private videotex system A system using videotex technology, installed by an organisation for private use. Some organisations have installed such systems to distribute corporate information internally, or to make their latest product information available to salesmen or customers.

private wire See *leased line.*

problem domain The field of application for an expert system. The problem domain defines the tasks, events and objects, together with their properties, that are represented in the knowledge base of an expert system. The current view is that, for the development of an expert system to be practical, the problem domain must fulfil these four main requirements:
  (1) Decisions must depend on a well-defined set of variables.
  (2) The values of those variables must be known.
  (3) The exact way in which decisions depend on the values of the variables must be known.
  (4) Interrelations among the variables in determining the decisions must be complex (otherwise conventional technology will be more cost-effective).

problem-oriented language A programming language designed to handle problems of a certain type. Thus ALGOL and FORTRAN are designed for mathematical problems, while COBOL is oriented towards business problems.
  Compare *declarative language*; *non-procedural language.*

procedural programming The commonest form of programming, in which programs are seen as a logical sequence of functions and procedures that operate on certain kinds of data. To develop a software system or an individual program, the complete logical sequence must be conceptualised, then the program procedure and the data must be designed to carry through that logic. Most of today's widely-used high-level languages, such as BASIC, COBOL or FORTRAN, are procedural.

Compare *declarative language*; *non-procedural language*; *object-oriented programming*.

procedure In general, a set of rules for achieving some defined purpose. Used more particularly in programming to mean the sequence of statements needed to solve a particular problem, as opposed to other elements of a program such as data declarations. Used more specifically still in some programming languages (PASCAL, for example) to mean a subprogram or subroutine that is similar in form to a complete program.

See also *block-structured language*; *stepwise refinement*.

procedure division The part of a COBOL program that contains the COBOL language statements intended to solve the processing problem.

process architecture See *application architecture*.

process control The use of computers to control industrial processes.

process decomposition diagram A diagramming technique used by systems analysts. A process decomposition diagram breaks down a business process, such as controlling stock, into the tasks involved in carrying it out. This process can be carried on to finer and finer levels of detail, forming an inverted tree structure.

processor The part of a computer that executes programs and controls peripheral transfers, roughly equivalent to the brain of a living organism. Sometimes referred to as the CPU, an abbreviation for 'central processing unit', it can be separated into three main components. In simplified terms, these are:

(1) a control unit that fetches instructions from memory and decodes them;

(2) an arithmetic/logic unit that carries out the processing those instructions specify, fetching data from memory for processing and storing the results there afterwards;

(3) various accumulators, registers and buffers, used to hold information in the course of processing.

Raw processor performance is usually measured in mips – millions of instructions per second. But a more accurate measure is achieved by running benchmarks, which are used to assess the performance of a processor for a particular workload.

processor cycle See *cycle time*.

processor limited Where the throughput capacity of a computer is determined by the speed of the processor.

Compare *input/output limited*.

profile See *functional profile*.

243

program A sequence of instructions or statements designed to tell a computer how to carry out a particular processing task. The term is used both to describe what a programmer produces (the source program), and also the instructions that are executed by the computer (the object program). Normally, special programs called compilers are used to translate the source program into object program form, and these contain a range of powerful features designed to make the programmer's job easier.

Sometimes called stored program, because the instructions are stored in the memory of a computer system before the program is executed.

program counter See *address buffer*.

program development The process of producing a correct, working version of a program. The main stages involved are design, coding and testing.

program documentation The information provided by the programmer to explain what a program does and how it has been constructed. It may include a description of the program; a flowchart; formats of inputs, data files, internal tables and outputs; and a listing of the source program. Its prime purpose is to help other programmers who may need to amend the program later, or to write further programs that interact with it.

program execution See *execute*.

program library See *software library*.

program listing A printed list of statements in a source program, produced by a compiler. As well as the original statements, it usually contains additional information designed to help the programmer, such as comments or codes identifying errors in the program syntax, or a map of the memory that the program will occupy.

program logic The way a program solves the problem that it sets out to address, or in other words the logical sequence of instructions or statements that it employs.

program maintenance The tasks associated with keeping programs up-to-date during their lifetime. This includes introducing and testing corrections or changes (arising because user requirements change); installing amended programs in the live system after testing; and updating program documentation.

program segment See *segment*.

program specification A statement of the requirements which a program is to meet. Normally a program specification is produced by a systems analyst or by a programming team leader and is used by a programmer as a basis for program development.

program status word (PSW) A word in computer memory, set by the operating system, that contains a number of indicators and codes reflecting the current status of a program that is running within the system.

program trace An aid to program debugging, either built into the compiler or interpreter, or carried out by a separate utility program. The trace records the sequence in which program instructions are executed so that logic errors can be identified.

programmable read-only memory (PROM) A form of semiconductor memory used to hold permanently-stored information. Unlike read-only memory (ROM) chips, whose contents are determined before fabrication, the contents of a PROM are determined after fabrication by a process known as blowing. A PROM is manufactured in the form of a blank which has tiny fusible wires in all bit positions. By applying a suitably high voltage in an appropriate pattern, the links that are not required are 'blown'.
  See also *EAROM*; *EPROM*.

programmer Someone who writes and tests programs. Most programmers do this as a full-time job and write programs to be used by other people, based on a program specification produced by a systems analyst or their team leader. But, with the arrival of personal computers, a number of people have become part-time programmers, writing programs for their own use or for their immediate colleagues, using programming languages such as BASIC.

programmer workbench A computer-based tool that helps programmers to produce documentation for the programs they design and write.

programmer/analyst A programmer who also undertakes systems analysis work for the programs that he or she develops.

programming The process of writing a program. It involves working out the steps necessary to solve the programming problem, then expressing these steps in programming language statements.

programming language A language used to write programs, either for computer systems or for other types of programmable machine. Languages for computers fall into two main categories:
  (1) procedural languages, such as COBOL, BASIC, FORTRAN;
  (2) non-procedural languages, such as report program generators or logic programming languages like PROLOG.

project As a verb, an operation (based on set theory) performed on a relational database that excerpts columns from one or more tables and uses them to produce a new table.
  See also *relational model*.

245

project champion An individual (from outside the information systems department) who champions – seeks to gain acceptance and support – for a development project. He or she will normally be in a position of influence in the organisation and may not be the main beneficiary of the project. The latter is likely to be the project sponsor – the user representative who sees the project through from adoption to installation. Some people believe that successful projects need both a project champion and a project sponsor.

project sponsor An individual, representing the intended users of a system, who accepts responsibility for the resources invested in a development project and sees it through from adoption to installation, perhaps chairing the steering group or acting as project manager. The project sponsor undertakes to deliver the benefits envisaged when a project is planned, whereas the development team of computer specialists undertakes to deliver the required hardware and software.

Prolog Contraction of PROgramming in LOGic, a high-level logic programming language used mainly for artificial intelligence applications, such as natural language understanding or expert systems. Prolog is a declarative language, rather than a procedural language like most other high-level languages. It is also a conversational language. Programming in Prolog consists of:

(1) declaring some facts about objects and their relationships (e.g. John likes Mary, Mary likes Jane),

(2) defining some rules about objects and their relationships (e.g. John likes anyone that Mary likes), and

(3) asking questions about objects and their relationships (e.g. does John like Jane?).

Prolog then works out the answer and shows it on the display.

PROM Abbreviation of *programmable read-only memory*.

prompt A short message or symbol presented on the display of a computer, indicating that it is ready to accept a new command or some other entry from the user. A well-known example is the '>' prompt used by some personal computer operating systems when awaiting a command.

propagation delay The time taken for a transmission block or a message to cross a network.

proportional spacing Used to describe a font or a line of text where every character is assigned its own width.

Compare *monospaced*.

protected field A field displayed on the screen (such as in a spreadsheet or in a 'fill-in-the-blanks' data entry format) that has been designated by the program displaying it as one that cannot be changed by the user.

246

protocol A set of procedures, usually formally specified, for exchanging information over a data communications network or over one of its components, such as a transmission line. Protocols for data communication have grown increasingly complex as their scope has increased, and they are defined both by leading computer manufacturers, such as IBM and DEC, and by international standards bodies, such as ISO and CCITT. See also (ISO's) OSI reference model, (IBM's) systems network architecture, (CCITT) V-series and X-series recommendations.

protocol conversion An operation by software which enables computer systems or terminals using different protocols to exchange information.

prototyping A way of developing an applications program or system by experiment. A first (prototype) version of the program or system is produced, usually missing some of the features and the polish of the final product. This is tried out by the user, with the programmer in attendance; then it is amended based on the experience; a new version is produced and tried; and so on until it is judged to be good enough to go into regular use. The technique is normally used with powerful system building tools that enable amendments to be introduced rapidly and with limited programming effort.

Sometimes, in what is known as throwaway prototyping, the prototype is discarded at the end of the prototyping process and a completely new version is constructed, probably using conventional development methods. The advantages of doing this are a more efficient and a more robust solution.

PSE Abbreviation of *packet-switching exchange*.

pseudo operation See *directive*.

pseudo real-time Used of a teleprocessing system that accepts transactions from online terminals in real-time, but then writes the transactions to a data file which is subsequently read back to update the master files. This is a safer and more efficient way of updating data files than real-time updating (in other words, updating immediately on receipt), but at the price of a delay (which may only be short) between the arrival of transactions and consequent updating.

PSS Abbreviation of *packet-switching service*.

PSTN Abbreviation of *public switched telephone network*.

PSW Abbreviation of *program status word*.

PTT Originally an abbreviation of post, telephone, telegraph, and still frequently used in Europe to refer to the national telecommunications authority, even though postal and telecommunications responsibilities have been separated in many countries. The term

247

National Telecommunications Agency (NTA) is also used.

public data network (PDN) A data communications network operated as a public service, by a national telecommunications authority or by a licensed private carrier. Public data networks are of two main types – packet-switched and circuit-switched (sometimes called fast circuit-switching). Generally they are national in coverage, but are interconnected so that international calls can be made.

public domain (PD) Non-proprietary, and thus not subject to copyright restrictions. Used particularly of software, often distributed by means of bulletin boards, and also known as *shareware*.

public key cryptography See *RSA*.

public switched telephone network (PSTN) The normal public telephone network.

publish and subscribe A dynamic equivalent of cut-and-paste – the mechanism for moving chunks of information around. Whereas cut-and-paste physically transfers the information as it is at the time the operation is carried out, publish-and-subscribe establishes a permanent logical link and transfers the information across whenever it is needed, such as when it is brought on to the screen or when it is printed. For example, a word processing document might 'subscribe' to a figure 'published' by a drawing application. Changes made to that drawing would subsequently be reflected in the document automatically.
    Compare *hot link*.

pull-down menu A menu that appears, by 'unrolling' down the screen, when activated by the user. Pull-down menus are part of the graphical user interface now available on many personal computers. Normally, only the menu titles are visible, along the top of the screen or window. When the user moves the pointer to a menu title with the mouse and presses the mouse button, the menu unrolls below it. The mouse is then used to move the pointer down the menu to the required option, and the mouse button is released to select it. The option flashes to show that it has been selected, then the menu disappears again.

pulse code dialling A form of dialling over telephone networks where the digits in the number are identified by a series of electrical pulses. Originally associated with finger-dialled telephones, but used also by push button telephones and modems on networks designed only for this type of dialling.
    Compare *touch tone dialling*.

pulse code modulation (PCM) A method of converting an analog signal such as a telephone call to digital form. The signal is sampled at intervals, and the amplitude is converted into a digital value. For speech signals the sampling rate is 8,000 samples per second, and eight bits are used to represent the amplitude. Thus digital speech requires a transmission bandwidth or bit rate of 64 kilobits per second. PCM techniques are used in conjunction with time division multiplexing to carry many speech signals along a single very high-speed digital channel. This is a very efficient form of multiplexing and it also eliminates the crosstalk experienced with multiplexed analog signals. At the receiving end, the signals are converted back to analog form and delivered to the telephone.

punched card A card used to represent data by means of small rectangular holes, now obsolete in computing terms. In the standard punched card, used both for data to be processed and for source programs, holes were punched, either singly or in combination, in 12 rows down the card and 80 columns across its length.

purge Remove unwanted or out-of-date information from storage or from memory, so that the space can be reallocated to other programs or users.

push Add an item to the top of a stack.
Compare *pop*.

push down list or **stack** See *stack*.

pushbutton dialling See *touchtone dialling*.

PVC Abbreviation of *permanent virtual circuit*.

Q

qualifier block A special block on a tape, preceded by a tape mark. Qualifier blocks are used to hold identifying information relating to a file or its contents, as opposed to the data itself.

quality assurance See *software quality assurance*.

quality of work life (QWL) A discipline concerned with the management and design of working procedures and systems that focuses on the quality of the working life of the people involved. The phrase was coined at a conference on the 'democratisation of work' held at Columbia University in the US in 1972.

query A request for a particular item of information from a data file or a database.

query language A formalised method of formulating a query and displaying the result on a screen. Query languages are provided as part of most database management systems, and enable data from the files to be located and displayed in a flexible way and without the delay and expense of writing a program expressly for the purpose.
  See also *structured query language*.

query processor See *query language*.

QWL Abbreviation of *quality of work life*.

R

---

ragged right Text that is not justified and so does not have a straight edge down its right hand margin.

RAID Abbreviation for 'redundant arrays of inexpensive disks', a storage technology for large-scale computer systems. Its basic principle is to employ a number of cheap disk drives such as are installed on personal computers, and to use these in parallel. The computer processor sees these disks as a single drive, but the controller is programmed to distribute data across the disks and/or to keep duplicate copies automatically, so that performance and reliability exceed that of conventional drives.

Five 'levels' of RAID have been defined. Level 1, for commercial applications needing high resilience, is based on disc mirroring; levels 2 and 3 are used mainly for scientific applications; while levels 4 and 5 are for commercial transaction processing applications.

RAM Abbreviation of *random access memory*.

RAM disk A form of cache memory often used on personal computers to improve performance. The RAM disk is an area of memory that is reserved to hold copies of program or data files (as specified by the user) that would otherwise have to be retrieved from disk when needed. Programs still access these files in the normal way for disk-based files but, since they are in fact in memory, access is much faster.

RAMP-C Abbreviation of *Requirements Approach for Measuring Performance – COBOL*, a benchmark for transaction processing systems developed by IBM. It uses a mix of four different types of transaction consisting of between 70 and 625 COBOL statements, and indicates transaction throughput when the system is loaded to 70 per cent of capacity.

random access memory (RAM) The most widely-used memory technology. In its simplest form, it consists of one transistor and one storage capacitor for each cell (ie bit) of memory. A charge in the capacitor indicates 1 and absence of charge 0. The transistor is used as a switch to connect the capacitor to a data line when the cell is selected for reading or writing. This is the cheapest form of RAM and is known as dynamic RAM or DRAM. Unfortunately, this type of memory loses its stored information every time it is read, and also by leakage from the capacitor. Therefore it requires its stored charge to be refreshed about once every two milliseconds, as well as after every read operation. Other designs, called static RAMs, do not require refreshing, but require extra transistors, which take up more space on the chip and result in a higher cost per bit. Today, 64 kilobit RAM chips are commonplace; 256 kilobit chips are being installed widely in the latest machines; and 1 megabit and larger chips are already in use.

random access storage See *direct access storage*.

random file A way of organising data files on direct access storage such as disk, sometimes known as hashed random or just hashed files. The block in which a record is placed is determined by applying an algorithm (known as the hashing algorithm) to the key value to generate a number within the range of blocks available for storage. If the block selected for a new record (the home block) is already full, the record may be written to an overflow block and a 'tag' recorded in the home block, consisting of the key value and a pointer to the overflow block. Alternatively, it may be recorded in the next sequential block that has space available. Records in random files can be accessed rapidly and efficiently, but efficiency declines in proportion to the number of records in overflow blocks, and, unlike indexed sequential files, they cannot easily be read in key sequence.

ras Abbreviation of *reliability availability serviceability*.

raster display A method for creating the image on a visual display screen. An electron beam continuously traces from left to right and from top to bottom across the display screen. Changes in the beam intensity determine what actually appears on the screen. This is the most widely used technology for visual displays, and similar to the technology used in ordinary television sets.

raster graphics Graphics composed of an array of picture elements (pixels), each of which is either white or black (or, in the case of grey shade or colour graphics, a grey shade or a colour respectively). Contrast with *object* or *vector graphics*, where the graphic image consists of a number of objects, each defined by its characteristics and position within the image.

raster image file format (RIFF) A format for files recording digital images, usually abbreviated to RIFF. Like the more popular but less efficient tagged image file format (TIFF), it records grey scale values at cell level, which means that images recorded in this format can be edited in detail on screen.

raster image processor (RIP) A device that enables typesetting machines to receive material from a personal computer for processing.

raster scan display See *raster display*.

rasteriser A software routine that turns a description of an object into a bit map that can be displayed or printed. For example, a rasteriser is used with outline fonts to convert each character into bit map form.

raw data Data before it has been processed and as a result rendered meaningful.

RBT Abbreviation of *remote batch terminal*.

RDBMS Abbreviation of *relational database management system*.

read-mostly memory See *EAROM*.

read-only memory (ROM) A form of semiconductor memory containing permanently stored information, such as program instructions or constant data values. A commonplace example is the program that controls a calculator or a digital watch, and most computer systems and terminals also contain one or more ROMs to hold frequently-used and invariant portions of the control software. Since the program or the data held in the ROM has to be fabricated on the chip, production cost is higher than for RAM chips, but performance is better. It also retains its contents when power is removed.

An alternative to fabricating a ready-made ROM is to manufacture a blank which has tiny fusible wires in all bit positions. By applying a suitably high voltage in an appropriate pattern, the links that are not required may be 'blown'. This type of memory is known as programmable read-only memory or PROM.

Not to be confused with CD-ROM, which is a storage device based on compact disk technology.

read time The time that an applications program needs to obtain a record from a direct access file. It includes both the time needed to transfer the record from the storage device, plus any time waiting for resources (such as a channel or the storage device itself) to become available.

read/write channel See *input/output channel*.

read/write head An electromagnetic component used in storage devices such as tape and disk to transfer information to and from the recording medium.

ready to send (RTS) A signal used to control data flow between two directly connected devices, such as a computer and a modem. The sending device raises the signal when it

253

has data to send. When it receives a clear to send (CTS) signal, it initiates a data transfer.

real address See *absolute address*.

real number A data type meaning a numeric field that may include digits to the right of the decimal point. It is normally held in floating point format within a computer.

real-time Used of a system where processing takes place immediately following the event that occasions it. Process control systems nearly always operate as real-time systems, because they must process the data arriving from the devices they are controlling quickly enough to feed back information affecting their operation. Teleprocessing systems such as airline reservation systems are sometimes described as real-time also, because they must record bookings immediately they arrive so as to maintain an up-to-date picture of the availability of seats. Both these types of system can be contrasted with batch processing.
See also *pseudo real-time*.

real-time clock A device within a computer system which generates signals at regular intervals of time. The real-time clock can be used by programs, via the operating system, to calculate the elapsed time between two events, or to cause themselves to be restarted automatically after a given time interval.
Compare *time-of-day clock*.

record The unit of storage for information held in a data file. The term is used in two senses – physical record and logical record. For storage on the recording medium, the file is divided into physical records, whose length suits the device concerned. For processing by applications programs or access by users, files are divided into logical records, each consisting of a regular pattern of one or more data fields. Each logical record normally represents one transaction or one entity. Thus a customer file would be divided into logical records each of which held the data relating to a particular customer. Each physical record holds one or more logical records.

record format How data fields are to be laid out in a data record. A record format identifies the different fields within the record, including key fields if any; shows their relative positions; and defines the data type and perhaps the permitted range of values for each field.

record length The specified length (for fixed-length records) or maximum length (for variable-length records) of records in a given data file.

record locking A means of controlling access to shared data files. Programs specify when they access a record whether they intend to update it. If so, the operating software prevents other programs from accessing the record until the update is complete. If the

record belongs to a database, then programs will also have to be prevented from accessing related records which may be affected by any changes, and this can be a complex
business. An effective means of locking records is essential to preserve the integrity of
shared files. Also referred to as *row locking*, a row in a relational file being the equivalent
of a record.

Compare *file locking*.

record overflow A condition that arises when data records in a direct access file will not
fit in the blocks which they should occupy based on the value of the key (the home block
or area). Consequently, they are stored somewhere else in the file where there is space.
Record overflow reduces processing efficiency because more accesses to the file are
needed to locate overflow records.

record update A modification to a data file consisting of the replacement of a record
with a modified version.

recording density The quantity of data that can be recorded in a given area of a storage
device. For tapes, this is normally defined in terms of bits or six-bit characters per linear
inch of tape. The recording (or packing) density of disks is usually expressed as the total
capacity of the disk in bytes. For example, floppy disks hold between about 100 kilobytes
and 1.5 megabytes, while the capacities of large disk drives and optical disks run into the
gigabytes.

records management The methods and procedures that serve to manage the records of
an organisation's operations, usually involving large volumes of information. Traditionally and still today, records management relies heavily on paper and, sometimes, micrographics filing systems. Computer technology has supported these systems by managing
indexes and retrieving records automatically (see computer assisted retrieval), but has
barely touched the records themselves. This situation is changing as more and more
records are generated by computer systems rather than originating on paper, and with the
arrival of optical disks capable of storing huge volumes of information at low cost.

recovery The process of restoring a computer system to a known and consistent state
after a failure, so that normal working can resume. The process is complicated because
the precise status of work that was under way at the time of failure usually cannot be
determined, since the contents of memory are lost. To avoid the danger that data files
have been left in an inconsistent state, the normal procedure is to reload an earlier version
of the files then bring them up to date – see *file recovery*.

recursive Where a program routine – a procedure or function – calls itself. For certain
types of problem, the use of recursion makes it possible to produce elegant and concise
program code which is more readable than the alternative – iterative code that deals with
the same problem step by step. The programming problem often quoted in the textbooks

as one where recursion is needed is 'The Towers of Hanoi'. This involves moving (a variable number of) wooden blocks decreasing in size from one pole to another, with a third pole available for temporary storage when needed. The rules are that only one block may move at a time, and a block may only rest on top of a larger block, never a smaller one. This problem is easily solved in a recursive high-level language like ALGOL or Pascal, but requires lengthy and tedious programming in a non-recursive language like BASIC.

reduced instruction set computer (RISC) A processor technology that uses a limited instruction set as a means of improving performance. A key design goal for these processors is to achieve one instruction execution per processor cycle, and thus simplify and streamline the processor logic. Computer design engineers believe that this outweighs the compensating disadvantage that programs will use more instructions to complete a given task than with a conventional (complex) instruction set.

redundancy (1) The inclusion, within a computer system or within its components, of additional capacity not required to meet normal demands. This additional capacity is brought into use when failures occur, thus achieving higher reliability.

(2) The attachment of extra information to a data field, so that its accuracy can be checked, for example after transmission or after being entered at a keyboard – see *cyclic redundancy check*.

reentrant Describes a program, or a section of a program, that does not modify its own logic in the course of processing. As a result, a teleprocessing system can use a single copy of the program to process a number of transactions concurrently, rather than needing to load a fresh copy for each – see *multithreading*. Some language compilers (notably COBOL) cannot easily generate code that satisfies this condition.

referential integrity A desirable condition for a database or for a set of related files, in which all cross-references between records are satisfied and there are no 'loose ends'. Ideally, referential integrity should be preserved automatically by software. Thus, for example, if the code identifying a customer is changed in the customer record, the software should find all the records of orders placed by that customer and replace the old customer code in those records with the amended code. Similarly, if a record is deleted the software should remove all references to it from other records.

refresh cycle A processor cycle which serves to refresh one of the system components, rather than to execute program instructions or to move data to and from memory. Some types of random access memory (RAM) and some types of display screen have to be refreshed periodically if they are to retain their contents.

refresh rate The frequency at which a component such as a display screen is refreshed.

register A component that a processor uses to hold data which has special properties for use during arithmetic or logical operations. Registers are normally the same length as the word length of the processor, or sometimes two words long. Computer systems may either have a number of general-purpose registers, addressed by the software as register 0, register 1, etc., or a number of specialised registers with symbolic names (A register, I register, etc.) each capable of particular functions, such as carrying out precision arithmetic or recording the results of tests and comparisons.

relation (1) In the relational model of database structure, a two-dimensional table – the basic format in which information is stored.

(2) For network or hierarchical models, a named association among sets of entities.

relational database management system (RDBMS) A database management system based on the relational model. This claim is often made (particularly for personal computer packages) principally on the grounds that the data is treated as a series of two-dimensional tables, known as relations. Stricter criteria would require also that algebraic operations, such as JOIN or PROJECT, could be used to manipulate the data and to create new tables based on various combinations of the original tables.

More detail *cost-based optimiser*; *referential integrity*; *two-phase commit*.

relational model A model for the structure of a database that treats the data as a series of two-dimensional tables (called relations) that can be manipulated using mathematical operations. The model was first defined in a paper by E. F. ('Ted') Codd of IBM, published in 1970.

A particular database based on the relational model is defined formally using mathematically-based rules which indicate precisely which tables should be included. This process is known as normalisation, and the resulting tables are said to be in third normal form. Data in third normal form is conceptually simple and can easily be updated without the risk of creating anomalies such as exists with more complex models of database structure, such as the widely-used network model.

Although conceptualised as a series of tables, formally a relational database is a mathematical set and can be manipulated by Boolean operations like AND, OR and NOT, or by algebraic operations such as JOIN (linking two tables to produce a third) or PROJECT (excerpt columns from one or more tables and use them to produce a new table). These features make the relational model highly flexible, and thus helpful for end-users wishing to retrieve and analyze data. By contrast, it is less helpful for computer specialists undertaking business analysis, since the tables must be translated into another form (such as by using the entity model) before the data can be depicted graphically. Databases constructed in terms of the relational model tend also to be relatively inefficient in their use of computer power.

Compare *hierarchical model*; *network model*.

relative address A number that identifies a position or a memory location within a specified area. Contrast with *absolute address*, which identifies an actual location. The absolute address is determined by adding the address of the beginning of the area – the *base address* – to the relative address.

release (1) Of a system resource, make available for use by other programs. A system resource, such as an area of memory or a peripheral unit, is allocated to a program by the operating system on request and is subsequently released by the program when it no longer needs it.

(2) Of a new version of a software or hardware component, or an amendment or upgrade, declare as available for use or purchase.

reliability availability serviceability (ras) The three qualities which contribute to good service from a computer system, originally coined by IBM and now with the status of a catchphrase.

relocatable Of a program or routine that can be moved around within memory and still executes correctly. This means, for example, that it must only address memory indirectly.
See also *indirect addressing*.

remote On another site. Contrast with *local*, meaning on the same site. Often used in a computercentric way, however, to mean 'on a different site from the main computer system'. Thus a so-called 'remote terminal' is in fact local from the point of view of the people using it.

remote batch terminal (RBT) A terminal, usually consisting of a keyboard and a high-speed printer, that is installed remotely from a computer system and can print the output from batch processing programs run on that system. This means that output can be produced at the location where it is required.

remote job entry (RJE) The entry via a terminal of control information and data to run a (usually batch) applications program on a remote computer system.

remote printing Where the output from an applications program is sent electronically to remote points to be printed, rather than being printed locally and forwarded by mail or some other means.

remote procedure call (RPC) A standard method that one application can use to ask another application, located in a remote computer, to carry out a specified task and return the results. For example, an application handling a user's query for information, finding that some of the data needed to satisfy the query was stored on another computer, could use a remote procedure call to obtain that data, then return a complete reply to the user's query. (The calling computer is known as the 'client', and the called computer the 'server'

– any type of computer may act in either of these roles.)

Thus the remote procedure call (often abbreviated to RPC) enables developers to extend an application over a network of computers, each of which has its own data files and user terminals but can also obtain data from the others when necessary. The RPC is a key part of the POSIX standard for a portable operating system.

remote slide projection Allows the participants in a teleconferencing session to view a series of predetermined visuals at each site, and sometimes to recall them as needed on a random basis. A conference leader at one site guides all the participants through the presentation.

remote terminal A terminal that is installed at a different location to the one occupied by the computer system that controls it, and connected by a communications link. Contrast with a *local terminal*, that is on the same site as the controlling system and connected to it by simpler (and cheaper) wiring and equipment.

remote testing Program testing where the programmer is not present when the test is run. The programmer supplies the program, test data and instructions as to how the test is to be run. Computer operators submit the test to the computer system and afterwards return the results to the programmer for analysis.

removable disk See *exchangeable disk*.

render Rendering software derives realistic three-dimensional images from a description of a scene and represents them on a computer display. It is used to create animation sequences, and also makes it possible for creative professionals such as architects or product designers to visualise ideas without having to build expensive models.

More detail *RIB*.

REPEAT statement A control structure used in some high-level languages to construct a loop within a program. The statement takes the general form, 'REPEAT action UNTIL condition', indicating that the statements constituting 'action' are to be repeated until 'condition' becomes true.

repeater A piece of equipment used to continue or extend a transmission line. It reconstructs and amplifies signals received from one part of the line, removing noise and distortion, then retransmits them on the other part. Repeaters are used at intervals in long-distance telephone lines, and can also be used to extend the length of a local area network or to increase the number of devices that can be connected to it. In terms of the OSI reference model, repeaters operate at level 1, the physical layer.

Compare *bridge*; *gateway*; *router*.

repetitive strain injury (RSI) A term used by ergonomists and medical staff to refer to physical disorders resulting from prolonged intensive work in awkward or constrained postures, such as by using a computer keyboard at a badly designed workplace.

See also *work-related upper limb disorders*.

report generator A program that can be used to produce printed reports, by specifying how the report is to be sequenced, where fields are to be placed within the print format, field totalling and subtotalling, and so on. Report generators are often associated with a database management system, operating on the data files that it manages.

See also *RPG*, a programming language originated by IBM.

repository An active and extended data dictionary. As well as details of the attributes of data items used by applications programs normally held in a data dictionary, it also records many other aspects of the way applications programs operate, such as screen formats, the contents of menus, business practices and other standardised procedures. It is described as active because it is accessed directly by the system building tools used to develop the programs, rather than merely serving as a documentary record.

requirements analysis See *feasibility study*.

requirements specification See *functional specification*.

reserved word A word that may not be used as an identifier by a programmer because it has been assigned a special meaning within the programming language concerned. Thus, for example, 'LET', 'FOR' and 'DIM' are reserved words in BASIC and therefore may not be used as identifiers for variables.

reset (1) Set a computer system into a predefined initial state. This is normally done by pressing a reset button, and is necessary to restart a system after a hardware failure or a program error that has halted or corrupted the operating system.

(2) In programming, set a count or an indicator to a normal or initial state.

resilience The ability of a system to absorb failures.

resolution The picture quality of a video image or of a visual display screen. The resolution of transmitted video signals such as television pictures is measured in lines per frame, and that of digitally encoded images in dots or pixels per inch. Document scanners normally operate at a resolution of 200-300 dots per inch, while higher resolutions are used for specialised applications such as mapping or pattern recognition. Lower resolutions (between 60 and 80 dots per inch) are used for standard visual displays, and their resolution is often expressed as the number of pixels that make up the horizontal and vertical dimensions of the screen. A typical personal computer screen has 640 horizontal by 480 vertical pixels, but the resolution can vary from 200-300 pixels for budget-quality graphics up to over 1000 pixels for high-quality screens.

See also *display size*.

resolution enhancement Techniques used by high-quality printers such as laser printers to enhance the appearance of the printed output. Some, for example, vary the size of the dots used around the edge of letters to sharpen up the outline.

resource (1) Any part of a computer system that a program can request the use of, such as an area of memory, a printer or a data file. Normally the program must reserve the resource first, either implicitly when it is loaded or explicitly by issuing a request to the operating system.

(2) On some computer systems, the term is used in a special sense to describe anything used by a program that is likely to change depending on the use made of the program. This is defined as a resource external to the program, rather than being included within it, so that it can easily be changed when needed. For example, an applications package designed to be used in a number of different national markets would include all the messages displayed on the screen and all the entries in menus as resources, calling them into the program when needed. To localise the English version of the package for, say, the French market, the English language set of resources is simply replaced with a new set in French, and the programs that make up the package do not need to be changed at all.

resource-sharing network A communications network used by computer systems at different geographical locations (commonly universities and other research institutions) to share one another's resources. The first resource-sharing network was ARPAnet, established in the early 1970s by the US Government's Advanced Research Projects Agency.

response frame A frame in a videotex database which allows the user to return information to the system, rather than just to retrieve information. Response frames are used, for example, to enable users to place orders for products.

response time The time a user, a terminal or a computer has to wait for a reply to a message. Most often used of teleprocessing terminals, where the response time includes the time for transmission in each direction, plus the turnaround time in the computer running the teleprocessing application.

restore Set back to a previous value or condition.

retention period The time in days which must elapse before a file may be overwritten. The retention period is held in the label or the directory entry of a data file, together with the date on which it was written. The operating system checks that the retention period has expired before it allows a program to overwrite or delete the file.

return address (1) The address to which control must be returned on completion of a subroutine.

(2) The address to which a reply to a message should be sent.

reverse out Put white text on a black or coloured background.

reverse video A way of drawing attention to a particular area of a visual display screen, by substituting black for white and vice versa.

revisable text format (RTF) A format for information representing a document that includes details of format as well as the content of the text. This would include, for example, the size of margins, the position of tabs, and formatting effects within the text itself, such as boldface and underline. The use of a revisable text format means that documents can be exchanged between different computers and different word processing packages and still assume exactly the same appearance on the screen.

Compare *text only format*.

rewritable optical storage See *erasable optical storage*.

RFT See *DCA-RFT*.

RGB monitor Abbreviation of red-green-blue monitor, meaning a colour display. Red, green and blue are the three colours used to build up a colour television picture, each produced by one of the three guns that drive the cathode ray tube.

RGB signal A signal containing colour (red-green-blue) information that can be used directly by a cathode ray tube, bypassing any television circuitry. This provides a clearer display than a composite video signal, which in turn is clearer than the broadcast radio frequency signal.

RIB RIB (standing for *RenderMan Interface ByteStream*) is a proprietary format for files describing three-dimensional images, defined by Pixar and incorporated into the company's RenderMan package. The format is supported by a number of rendering and modelling packages.

It incorporates a programming language that is used to describe the geometry of a three-dimensional image, much as page description languages such as PostScript are used to describe two-dimensional images. The language uses shaders – subfiles describing the attributes of a surface or object – that can be applied to parts of the image much as fonts are used by PostScript.

See also *IGES*.

rich text format See *revisable text format*.

RIFF Abbreviation of *raster image file format*.

right justify See *justify*.

ring network See *ring topology*.

ring topology A topology, used particularly for local area networks, in which the transmission medium runs from device to device in a ring. Signals travel round the ring to each device in turn. Normally, the devices in the ring can detect if a device or a segment of the ring fails and arrange to bypass the failure automatically. Ring networks normally use a token-passing protocol to control traffic. See, for example, *Cambridge Ring*.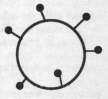

RIP Abbreviation of *raster image processor*.

RISC Abbreviation of *reduced instruction set computer*.

river A white space running down through the middle of a column of justified text. It spoils the appearance of text and can be cured by adjusting hyphenation or point size.

RJE Abbreviation of *remote job entry*.

robot A programmable or 'teachable' device, roughly on a human scale, and with some ability to sense its environment.

robotics The technologies on which industrial and other robots are based.

roll call polling A method of polling terminals (see *poll/select*). The controlling device invites each terminal on a line in a predefined sequence to send or receive information.

roll in Bring (a program or part of a program) into memory from temporary storage on a direct access storage device such as disk. To make more efficient use of the memory that is available, a multiprogramming operating system may write programs that are waiting to execute to direct access storage (in other words, roll them out) in order to release the memory they occupy for the use of other programs. When the turn of one of these 'rolled out' programs arrives to execute, the operating system rolls it in again.

roll out Store (a program or part of a program) temporarily on a direct access storage device such as disk.
　　Compare *roll in*.

ROM Abbreviation of *read-only memory*.

router A component of a switched data network that decides which route messages or blocks should follow through the network. It may either be a discrete hardware component or a software routine contained in all of the network nodes. It examines the address information in each incoming message or block in turn, decides by which route it should continue its journey, then forwards it on the chosen outgoing line. Routing may be determined by the directness or cost of alternative routes and/or by priorities set by the operator of the network.

In terms of the OSI reference model, routers operate at level 4, the network layer.

Compare *bridge*; *brouter*; *gateway*.

routine A section of a program that carries out a specified task.

routine maintenance See *preventive maintenance*.

routing page See *index page*.

row A record in a relational file.

row locking See *record locking*.

RPC Abbreviation of *remote procedure call*.

RPG Abbreviation of *Report Program Generator*. A non-procedural programming language used, as its name suggests, to write programs to produce printed reports.

See also *report generator*.

RS-232 A standard defining the electrical interface for serial communication between small computers and peripherals such as printers and modems. Functionally identical to the CCITT V.24 standard.

RS series recommendations A series of standards recommendations specified by the Electrical Industries Association (EIA) of the US. RS is an abbreviation for requirement specification. The best known of these standards is RS-232.

RSA An encryption method invented by (and named after) Rivest, Shamir and Adelman and licensed by a number of major computer and software manufacturers for incorporation into their products. It uses a technique called public key cryptography. Rather than using the same encryption key to both encrypt and decode messages, it uses a matched pair of keys, one of which is made public while the other is private. The sender looks up the public key to encrypt a message, and then only the recipient's private key can decode it. This makes management of keys simpler than with the other widely-used method, data encryption standard (DES). For this reason RSA tends to be used over public networks between independent organisations, while DES is popular for single-organisation private networks, such as a bank's funds transfer network. RSA can be used independently or in conjunction with another encoding technology such as DES.

RSI Abbreviation of *repetitive strain injury.*

RTF Abbreviation of *revisable text format.*

RTS Abbreviation of *ready to send.*

rule A method of expressing knowledge to be included in a knowledge base. A rule consists of a number of facts that have an IF...THEN relationship, and may also include the degree of certainty (the confidence factor) that can be attached to any conclusion drawn from them.

rule template See *graphical rule language.*

run One performance of a program or routine, by starting it and allowing it to continue to its conclusion. As a verb, a synonym of execute.

run length encoding A data compression technique, in other words a method of reducing the number of bits used to represent information, such as for facsimile or modem transmission or for storage as document images. The length of continuous sequences of 0s or 1s is measured and recorded, rather than recording the sequence itself.

    See also *Huffman encoding*; *LZW algorithm.*

run time The period of time when a program is executing. The term is also used to describe actions or decisions taken at this stage rather than, for example, when the program is being compiled – known as *compile time.*

S

S-100 bus A type of bus designed for personal computers. For early personal computers it acquired the status of a *de facto* standard, and was adopted by a number of suppliers.

S-video A type of component video signal generated by more expensive home VCRs.

SAA Abbreviation of *systems applications architecture*.

sampling rate The number of times in a given period, normally a second, that the value of an analog signal is measured, in order to convert it into digital form.
    See also *pulse code modulation*.

sans serif See *serif*.

satellite processor A processor which forms part of a larger computer system and whose function is to run programs at the dictates of the processor controlling the system as a whole. Normally, a satellite processor is installed at a different location from the controlling processor, and is used where it is advantageous to carry out some processing at that location, rather than process centrally and transmit or transport the results.

SatStream British Telecom's trade name for digital transmission services via satellite. It is intended for long-distance international transmission with speeds up to about 2 megabits per second.

saturation testing Testing designed to discover how a system performs under conditions of heavy load, such as when, for example, all terminals attached to a teleprocessing system attempt to enter messages at the same time.

save Store (a data file or a program or the contents of memory) semi-permanently on mass storage, such as disk or tape.

266

SBT Abbreviation of *system building tools*.

scalar type A data type that can have any of an ordered set of values that are listed when it is declared. For example a scalar type called 'weekdays' could be declared as having the values 'Monday', 'Tuesday', 'Wednesday', 'Thursday' and 'Friday'.

scaleable processor architecture (SPARC) A flexible architecture for small processors such as graphics workstations, originated by the US manufacturer Sun.

scanner A device which reads printed or written material and converts what it finds into a digital form, so that it can be transmitted or be processed by computer. The most familiar example is the fax machine, and five other main types of scanner are used in association with computers:

(1) *Document scanners* capture pages of text or drawings, and are used for desktop publishing and document image processing.

(2) *Bar code scanners* read the bar codes used to identify products, as at supermarket checkouts.

(3) *Mark readers* or scanners detect handwritten marks on preprinted forms, and are used for such applications as meter reading and computer-marked tests.

(4) *Magnetic ink character recognition* (MICR) readers are used to process cheques.

(5) *Optical character recognition* (OCR) readers recognise printed characters and are used to capture documents for word processing or for information retrieval.

scene understanding See *machine vision*.

scheduled maintenance See *preventive maintenance*.

schema An expression in programming terms of a data model, in other words a programmed description of items of data that are to be included in a database, also showing the relationships between them. There are three different types of schema, each representing a different viewpoint on the data:

(1) The *conceptual* schema describes the data in business terms, independently of the means used to store it.

(2) An *external* schema (sometimes called sub-schema) is a subset of a conceptual schema representing the view of the data taken by a particular applications program or user.

(3) An *internal* schema (sometimes called a storage schema) represents the data as it is physically stored on the mass storage devices. See also *data independence*.

scientific data processing Data processing in support of scientific and technical work. Timesharing services are widely used for this purpose. It sometimes requires computer systems with great computational power, sometimes referred to as supercomputers. Contrast with *commercial data processing*, where the main requirement is the ability to

267

handle large data files held on mass storage and networks of data terminals.

scratch tape A tape that is available for use, such as to hold new information or to be used as a work tape.

scratchpad memory See *work area*.

screen buffer An area of computer memory from which a display reads the information to be displayed on the screen.
See also *bit-mapped display*.

screen format How information is to be arranged on a display screen used for a particular purpose, such as to enter a particular type of transaction. The format shows the fixed descriptive text or other information provided to guide the user, and identifies the areas on the screen into which data may be entered. Where a mouse or a touch-sensitive screen is used, the format may also include control fields such as buttons or pull-down menus.

screen frequency The spacing of the black spots used to create a halftone image, usually measured in spots per inch. At a given resolution, a higher screen frequency produces a more refined looking image, but with fewer shades of grey and therefore more noticeable jumps from one grey level to another.
See also *posterise*.

screen layout See *screen format*.

screen mode Where the screen of a visual display device is treated as a fixed space, that can be replaced completely by a fresh display of information or be changed by overwriting parts of it.
Compare *scroll mode*.

screen painter A program, usually associated with a database management system or an application generator package, that can be used to describe the format of display screens used, for example, to enter data or queries. It allows data fields and descriptive text to be placed wherever required on the screen.

screened cable A cable in which the conductor is surrounded by an earthed metal screen, to reduce interference from outside electrical signals.

screening See *halftoning*.

script A section of program code associated with a particular event or condition, such as a value keyed into a particular field displayed on the screen or a mouse click on a button. When the event or condition occurs, the script is executed and takes whatever action is appropriate to react to it.

scroll Move the image on a computer screen up or down.

scroll bar A narrow rectangular box alongside or beneath a window on a computer screen, which is used to position the image displayed within the window. The scroll bar is operated with a mouse by placing the pointer in different parts of the bar and pressing the mouse button. Normally, scroll bars have arrows that scroll the image in the direction indicated, and a sliding box (the *scroll box*) that can be moved along the bar to position the image rapidly, such as when looking for a particular page in a large document or a particular cell in a spreadsheet.

scroll box A movable box within a scroll bar. Its position within the scroll bar indicates the relative position of what is currently displayed, such as the position of the current page within an entire document.

scroll mode Where the screen of a visual display is treated like a printer with a continuous roll of paper. Information is displayed a line at a time, working down the screen. When the screen is full, lines begin to disappear off the top of the screen and all the other lines move up behind them.
   Compare *screen mode*.

SCSI Abbreviation of *small computer systems interface*.

SDH Abbreviation of *synchronous digital hierarchy*.

SDLC Abbreviation of *synchronous data link control*.

SECAM Abbreviation of sequentiel couleur a memoire, a standard for colour television broadcasting adopted in France, the Soviet Union and elsewhere, with 625 picture lines and a field frequency of 50Hz. Contrast with *PAL*, used elsewhere in Europe including the UK, and *NTSC*, used in the US.

second generation language See *assembly language*.

second sourcing An agreement by which the original designer of a particular integrated circuit licenses other manufacturers to produce the same design.

secondary memory See *memory*.

sector See *disk sector*.

security The ability of a computer system or a computer installation to resist threats to its continued operation, such as from human error, accidents or criminal acts. Security of access to information held on a computer system is usually referred to as privacy.

seek area See *cylinder*.

seek time The time taken by a disk drive, or any similar direct access device, to position the read/write heads over a given track. Sometimes expressed as the average time needed to reach any track.

segment A section into which a large program is divided, either to simplify programming or to limit the amount of memory it needs to run. In the latter case, segments of the program are only loaded into memory when needed.
See also *overlay*.

select (1) Indicate to a terminal on a multipoint line that it may use the transmission line – see *poll/select*.
(2) Indicate to a peripheral that it may use a selector channel to transfer data to or from memory.

selector channel A channel within a computer system that can be 'selected' by high-speed peripherals such as disk drives, to transfer a burst of data (such as a physical record) between the peripheral and memory. A number of such peripherals normally share a selector channel, and compete to select it whenever they have data to transfer.
Compare *multiplexor channel*.

semantic data model In general, a model of the structure of a set of data that seeks to describe the real-world characteristics of the data, avoiding computer-related terms. Used specifically to mean an extended form of entity-relationship model that expresses additional real-world subtleties (hence also called the enhanced entity-relationship model). For example, entities may be organised into classes or superclasses in which some but not all of their attributes are shared; may be divided into subtypes; or combined into composite objects.
More detail *entity subtype*; *class hierarchy*.

semantic network A method of representing knowledge. The information is represented as a set of labelled nodes, representing objects or ideas, connected by labelled arcs or lines that represent the relationships between them.

semiconductor The base material used to fabricate transistors and integrated circuits. A semiconductor is a piece of pure crystalline material – usually silicon, obtained from sand – into which tiny amounts of other materials have been introduced.
See also *dopant*.

semiconductor memory See *random access memory*.

sense line A wire in the connection between a computer and a display screen that the computer uses to determine the resolution at which the screen is to operate.

separation A term used in desktop publishing to refer to the different partial versions of a page or an image that the software will produce. If the material is to be colour printed, for example, separations will be produced for each of the colours to be used in the printing process.
See also *CMYK*; *RGB signal*.

separation table A table of information used to convert a particular colour into the separate colour values used to reproduce it, such as the CMYK values widely used for printing.

SEQUEL See *structured query language*.

sequential Arranged in a predetermined sequence, based on one or more key fields. Data files are organised in this way so that they can be processed in a known sequence –transaction or amendment files are sorted into the same sequence as master files then are read in parallel with them. Sequential files on direct access storage devices often also have an index so that individual records can be located directly – see *indexed sequential*.

sequential processing Processing of data in a predetermined sequence, such as the sequence of the keys on the master files. Batch processing applications frequently use sequential processing, sorting transaction files into sequence before processing them against master files.
Compare *transaction processing*.

serial file A data file from which records can only be read in the same sequence that they were written originally. Files on magnetic tape are necessarily serial because of the nature of the medium, but serial files may also be used on direct access storage where there is no need to read records out of their original sequence, such as for a print file or a transaction journal.
Compare *indexed sequential file*; *random file*.

serial interface A method of transferring data between a peripheral and the memory of a computer system, in which bits are transferred one after another. Also used to refer to the device which executes the transfer.
Compare *parallel interface*.

serial storage Storage where records can only be retrieved in the same sequence that they were written originally, such as magnetic tape. Contrast with *direct access storage*,

271

where records can be retrieved independently of the sequence in which they were first stored.

serial transmission See *serial interface*.

serif An end stroke on the head or tail of a letter, used in certain typefaces such as Times or Bookman. Serifed typefaces are easy to read. Sans serif typefaces like Helvetica, by contrast, have no such strokes and have a more authoritative look.

server Generally used to mean a computer system, attached to a communications network, that provides a particular service to other devices on the same network (known as its clients), on demand. It is also used to mean a program running within a computer system that provides a service to other programs in that system in a similar way.

A server is identified according to the service it provides. Thus, for example, a local area network might have a print server attached to it, equipped with a couple of laser printers. When required, personal computers attached to the network would send documents to the print server to be printed. Servers are used to enable a number of otherwise independent computer systems to share a relatively expensive resource, such as the laser printers mentioned above; to access a shared resource such as a file of common information (a file server); or as a common point of access to other resources outside the network, such as a public data network or a remote mainframe computer (a communications server, otherwise known as a gateway).

The term is also used in almost the reverse sense in the context of the X Windows networking protocol, to mean a device with a screen and keyboard that is controlled by an application running in another computer attached to it via a local area network (the client).

See also *client-server*.

service bureau A business organisation that hires out computer time. Most service bureaux hire out time on particular applications packages and on timesharing services, rather than raw computer power for customers to use as they wish.

service level agreement (SLA) An agreement drawn up between the supplier of a service (such as an in-house information systems department or an external contractor) and its recipient (such as a departmental manager) defining the performance levels that the supplier will be expected to meet. It might specify, for example, that transactions will be processed with given average and maximum response times; that processing services will be available for a given percentage of the working day and will never be unavailable for more than a given time; and so on.

service provider A business organisation that operates a service accessed by customers via a communications network, such as a videotex service. This is in distinction to the *information provider* – the organisation that owns the information that the service provider distributes.

session Of the use of a data terminal, between one log on and the subsequent log off. During a session, a transmission path is normally held open by the network, and processing resources are held available by the computer system that is being used. Many transactions or exchanges of messages may occur in one session.

session layer The fifth protocol layer of the OSI reference model, defined by ISO. The session layer establishes and manages sessions between users' terminals and applications programs or other services.

See also *application layer*; *data link layer*; *network layer*; *physical layer*; *presentation layer*; *transport layer*.

session level See *session layer*.

set (1) In the relational model of database structure, a collection of things – the normal mathematical definition.

(2) In the network or hierarchical models, a named association of a number of entities. A set of this kind is defined on the assumption that applications will normally want to process all the records in the set as a group rather than deal with them individually.

SGML Abbreviation of *standard general markup language*.

shadowing A phenomenon apparent on some types of display screen, and noticeably the LCD (liquid crystal display) screens used on many portable computers. Also known as *ghosting*. Because it takes time to change the colour of a pixel, moving objects on the screen leave a ghost trail. Vertical and horizontal shadows may also spread out from large dark regions or rectangular objects because the current powering the display 'leaks' into neighbouring cells

shallow knowledge Knowledge developed through training and experience, commonly referred to as expertise. This is in contrast with *deep knowledge*, which is more general, theoretical knowledge. The terms are used in the expert systems field.

shared filing See *electronic filing*.

shared logic An arrangement in which a number of terminals share a single, local processor running the application to which they are all dedicated. Used particularly to describe the shared-logic word processing systems installed in many typing pools in the late 1970s and early 1980s (as opposed to the standalone word processors used by individual secretaries). These had a number of visual display units each of which could work separately on different documents. Documents were stored on a shared hard disk and printed on one or more shared printers.

shareware Software, mainly for personal computers, that is distributed outside the normal commercial distribution channels, via bulletin boards and by direct exchange between users. Either this is free software, (known as public domain software) developed by non-profit institutions like universities or public-spirited individuals, or alternatively the developer relies on the honesty of users. Such software displays a message asking users to send a fee (usually in the range $5 to $50 – much less than normal retail prices) to the developer if they decide to keep the software, or otherwise to stop using it or pass it on to someone else. The message usually points out that the best way to make sure that you continue to get good value shareware is to be honest and pay up. This is also known as *honorware*.

sheet feeder An attachment to a printer that enables individual sheets of paper, rather than continuous stationery, to be fed through it.

shell A framework that serves as a starting point for an application or a program that is built to fit within it.
     See also *expert system shell; Unix.*

short haul modem A simplified and low cost modem that can only transmit data over relatively short distances. Normally used for in-plant applications.

sign bit A bit, normally at the lefthand end of a single or double-word value, that indicates whether the value is positive or negative. Normally, negative numbers are represented within computer systems as (ones) complements of the corresponding positive number, so that a 0 sign bit means positive and a 1 means negative.

sign off See *log off.*

sign on See *log on.*

signalling The exchange of information between the components making up a communications network, in order to control the handling of information exchanged by the network's users. Signalling is used particularly to establish and clear down calls.

silicon chip See *integrated circuit.*

silicon drive See *memory card.*

Silicon Glen See *Silicon Valley.*

Silicon Valley An area of California in which a number of innovative information technology companies are based, and from which in particular the first personal computers emerged. By analogy, the area of Scotland that has attracted a number of information technology manufacturers has been dubbed 'Silicon Glen'.

SIMD Abbreviation of *single instruction multiple data*.

SIMM Abbreviation of *single in-line memory module*.

simplex Of a transmission link where transmission can take place in only one direction. Compare *full-duplex*; *half-duplex*.

Simula A programming language designed to handle simulation problems and regarded as the forerunner of object-oriented programming.

simulate Operate in the same manner as, in terms of the activities undertaken. This is normally achieved via software programmed to behave as nearly as possible like the device being simulated. Aircraft simulators are the best known example, providing on the ground the nearest possible representation of conditions in an operating aircraft. Simulators are also used within computing to enable new types of computer to continue to process the workload of the machine they have superceded. Long after punched card machines had disappeared, for example, many organisations were still running some of their old card-based applications via simulators, pending the time when these could be re-written.
Compare *emulate*.

single in-line memory module (SIMM) A small printed circuit board containing a certain amount of random access memory, that can be plugged into standard sockets in the motherboard of a personal computer. These components normally hold 8 memory chips making up 256 or 1024 kilobytes of memory in total, and can easily be added or exchanged to extend the memory capacity of the computer.

single instruction multiple data (SIMD) A type of parallel processing machine that performs the same operation on many different items of data at the same time.
Compare *multiple instruction multiple data*.

single mode (SM) Used of optical fibre that only allows a single ray of light along the fibre. It is relatively small, with a core only 8 microns in diameter. Contrast with *multimode* fibre, which is cheaper but does not perform as well. Also known as *monomode*.

single-sided Used of floppy disks where only one surface of the disk is suitable for holding data.

single step operation Where a program is executed one instruction at a time. This mode of operation may be used by engineers to trace hardware faults or by programmers to trace obscure faults in operating software.

site licence A licence issued by the supplier of a software package to permit a customer organisation to use the package anywhere on a specified site. To protect their copyright in the software they have developed, most suppliers of software packages impose terms on the purchaser restricting use to the purchaser personally. Some also offer site licences to large organisations that may require many copies of a popular personal computer package, at a bulk discount.

site network The part of a network that connects user devices such as computers and terminals to one another and to switching systems, rather than connecting together the sites and switching systems. The latter is termed the backbone or trunk network.

SITPRO Abbreviation of *Simplification of International Trade Procedures Board*, an organisation set up by the British Overseas Trade Board. In the late 1970s SITPRO developed a set of syntax rules for exchange of trading data between business organisations. These rules were adopted as guidelines for the development of later standards for electronic data interchange, such as Tradacoms and Edifact.

16-bit (1) Used of a microprocessor or a bus (the data 'highway' within a small computer) to specify how many bits it can handle at once. Together with speed of operation (expressed in MHz) this is an important indicator of performance.

(2) Also used to specify how many bits are used to achieve a particular effect, as in '16-bit colour' or '16-bit addressing'. Since 16 bits can carry 2^{16}, i.e. roughly 65,000, combinations, a computer using 16-bit addressing is capable of addressing directly that number of memory locations.

sizing The process of deciding the hardware capacity required to support a given application – processor power, quantity of memory, number of channels, number of disk drives, and so on. For an application of any size and complexity, accurate sizing is an important and difficult exercise.

SLA Abbreviation of *service level agreement*.

slave memory A section of memory used independently by one component of a high-performance processor (such as a pipeline processor) or by a peripheral controller.

slave store See *slave memory*.

sliding window A mechanism used in communications protocols to achieve high throughput. With simple protocols, the sending system waits after sending each message until it receives a positive (ACK) or negative (NAK) response from the recipient. With a sliding window protocol, by contrast, a number of messages can be sent without waiting, and acknowledgements refer to all the messages received up to a specified point.

slotted ring A type of local area network with a ring topology in which devices using the network share access through predefined, short message slots. It originated from the Cambridge Ring network and is defined by ISO standard 8802/7.

slow scan video See *freeze frame video*.

SM Abbreviation of *single mode*.

small computer systems interface (SCSI) A standard for the interface used to connect high speed peripherals such as hard disks to personal computers and other small computer systems. Often abbreviated to SCSI, pronounced 'skuzzy'.

small scale integration (SSI) Integrated circuit technology used for fairly simple circuits involving a hundred or so logic elements.

Smalltalk An object-oriented development system. It developed out of the work of Alan Kay at the University of Utah in the early 1970s and subsequently at Xerox Corporation's Palo Alto Research Centre (PARC). Smalltalk was associated with the development at PARC of the first graphical user interface, itself an object-oriented software environment, and that was later embodied in the Apple Macintosh personal computer.

smart building See *intelligent building*.

smart card A plastic card, similar in size to a credit card, into which an integrated circuit has been embedded, so that the card can record and process a small amount of information. Smart cards are considerably more expensive than the magnetic stripe cards widely used as credit cards and cash dispenser cards. Magnetic stripe cards cannot themselves process information, but can be read from and written to by the machines into which they are inserted. By contrast, smart cards can be made virtually fraudproof, and can also hold much more information. Smart cards are in wide use in France, the US and Japan. They are being used for electronic payment systems, as telephone charge cards and to hold patients' medical records.

smart quotes Automatic conversion by word processing software of the vertical quotation marks available on most keyboards into the "sixes and nines" quotation marks normally used in printed text.

smart terminal See *intelligent terminal*.

SMP See *TCP/IP*.

SMPTE Abbreviation for *Society of Motion Picture and Television Engineers*, originators of a standard for timecodes on video recordings. The format shows hours:minutes:seconds:frames, as in 11:39:15:05.

SNA Abbreviation of *systems network architecture*.

snapshot dump A record of selected areas of memory, printed out (or written to direct access storage for subsequent printing) at chosen points while a program is running. The programmer uses snapshot dumps to help debug programs under test.

sneakernet Humorous description for the exchange of data between personal computers by transferring floppy disks, by association with terms for local area networks such as Ethernet.

SNMP See *TCP/IP*.

snowflake topology A topology, used particularly for local area networks, in which devices are connected in stars, the central points of which are connected to a central hub for the whole network.

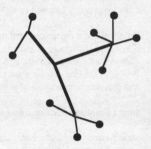

soft copy Output from a computer system that is displayed temporarily on a screen, rather than printed in a permanent form.

Compare *hard copy*.

soft key A key on a keyboard which has a different effect depending on the program in use.

soft sectoring Of floppy disks requiring only one positioning hole in the disk to enable the drive to locate the sectors on the disk. Contrast with *hard sectoring*, where a number of holes are punched in the disk.

soft systems methodology A methodology for dealing with messy, ill-structured problems, particularly associated with Peter Checkland of the University of Lancaster. It has been applied to general management problems as well as (in fact more than) to problems of information system design. Its salient quality is that it uses systems models (known as 'holons') as a means of thinking about a problem situation. It does not assume that the problem situation necessarily constitutes a system that can be engineered, and this distinguishes the approach from systems engineering (described as 'hard' systems thinking), which does proceed from that assumption.

software In general, software consists of all the components of a computer system that are programmed rather than manufactured, in other words, those that are not hardware. By contrast with *hardware*, the operation of software can easily be altered, by re-programming or by reconfiguring. Thus it serves to help people match computer systems to a particular set of requirements. Software components can be divided into three main categories – applications programs and applications packages that do the particular work required of a system; the system building tools and compilers that are used to help build the applications programs and packages; and the operating software, normally supplied by the computer system manufacturer, that controls the operation of the hardware and of the other software components that the system runs (such as those in the two other categories).
See also *firmware*.

software engineering A highly disciplined approach to developing software, regarding the production of software as an engineering problem. It uses rigorous technical methods, reinforced by rigorous management methods, and is practised mainly by computer manufacturers and software suppliers, and to develop military systems.

software house A business organisation that develops and installs applications programs or other types of software for its customers to run on their own computer systems.

software library A collection of standard routines, provided either by the computer supplier or by the user organisation or both, that can be called into programs automatically when needed.

software metrics Methods of scoring the various qualities of a software product, so that its quality can be measured. Software metrics are used within software quality assurance schemes.

software package A program, or a set of programs, designed to meet the requirements of a particular class of user, rather than those of an individual user.
See also *application package*.

software piracy The theft of the ideas incorporated in a piece of software, either by copying the software for others to use or by incorporating those ideas in another software product, without permission from the author or publisher.
See also *Federation against Software Theft*.

software publisher A business organisation that commissions, assembles and markets software products such as applications packages. It sells these via distributors and retailers and direct to customers.

279

software quality assurance (SQA) The process of ensuring that a software product is of acceptable quality or, in other words, that it conforms to the requirements established for it. Those requirements derive both from the needs of the intended user of the software and from general quality requirements, such as that the software should be easy to change or to maintain.

solid model A computer-generated three-dimensional representation of an object that can describe its physical properties, such as mass.
  Compare *wireframe model*.

sort Arrange the records in one or more data files into a predetermined sequence, based on the values in one or more of the data fields they contain. These data fields are known as key fields.
  Compare *merge*.

sort generator A program that can be used to create sorts and merges. The user specifies which data files are to be used for input and output, which fields are to be used as keys for the sort (or merge), and perhaps criteria for selecting records from input files.

sort program generator See *sort generator*.

sorting sequence See *collating sequence*.

soundex A function used in searches of text files, which matches words or parts of words according to their sound rather than their spelling.

source When a file or document is copied, the original as opposed to the duplicate version (the *destination*).

source code See *source program*.

source data capture Capture of data recording a transaction at the point where it originates, such as in a bank branch or at a supermarket checkout. The advantage of source data capture, using online terminals, is that data can be validated and corrected on the spot before it enters the processing cycle.

source document A document which records the original event or transaction that causes data to be input to a computer system for processing.

source language See *source program*.

source lines of code (LOC) Source lines of code per person day or per person month is the conventional way of measuring programmer productivity. Its limitation as a measure is that it disregards activities such as requirements analysis and systems design that precede programming in the development lifecycle, but may well be more important. Such measures also depend on the equipment or the programming language that is used, or on how the development team is organised. An alternative measure is now coming into use, called *function point analysis*, that does not share these defects.

source portability See *portable*.

source program A program in a programming language, as written by the programmer. To be run on a computer system, the source program must be translated into machine code form. This is done either by a compiler, which produces a new version of the program known as the object program, or by an interpreter, which reads the source program, translates it into machine code, and executes the code as it goes.

SPARC Abbreviation of *scaleable processor architecture*.

spare tracks Tracks reserved on a disk to substitute for any tracks found to be faulty. Disk drives or controllers automatically substitute a spare track for any track found to be faulty when a disk is initialised or when an attempt to write to the track fails.

SPC Abbreviation of *stored program control*.

specialisation See *inheritance*.

speech channel See *voice channel*.

speech processing See *voice recognition*.

speech recognition See *voice recognition*.

speech synthesis See *voice response*.

speech traffic See *voice traffic*.

spelling checker A piece of applications software, often built into a word processing package, that checks the spelling of words in a document and flags those that (it thinks) are misspelt. Spelling checkers normally compare words against a dictionary of correct spellings, allowing users to amend this or to add further spellings if they wish.

split screen Where a visual display screen is divided into (usually two) areas, each of which is treated as if it were a separate and independent screen. This is done, for example, to allow the user to interact with two applications at once, each of them using its own area of the split screen. Or, in word processing applications, it allows the user to look at

two different documents, or two parts of one document, at the same time.
See also *window*.

sponsor See *project sponsor*.

spooler A routine that reserves part of the memory or disk storage of a computer system to hold data en route to a slow output device such as a printer. After it has stored information in this way, the spooler allows the originator of the data (applications program or personal computer user) to continue with other work, and itself ensures that the printing is completed. Print spoolers on personal computers normally run in background mode.

spreadsheet An applications program (usually a package) that helps users to build up two-dimensional tables of numeric and text information. Spreadsheets are widely used for budgeting and financial modelling. As well as entering information, the user can also embed formulae into the various cells in a table, such as to compute column or row totals or work out variances. When fresh data is entered, the spreadsheet software automatically recalculates any other cells that are affected. The first spreadsheet package, called *Visicalc*, was responsible for many of the early sales of personal computers for business use in the late 1970s, but has since been overtaken in popularity by packages such as Lotus' *1-2-3* and Microsoft's *Excel*.

sprite A term used in computer-generated graphics to mean a collection of graphics that are used to create an animated sequence.

sprocket feed See *pin feed*.

sprocket holes The holes punched in the outer margins of continuous stationery. Rotating pins or sprockets engage in these holes to draw the paper through the printer.

SQA Abbreviation of *software quality assurance*.

SQL Abbreviation of *structured query language*.

SRAM Abbreviation of *static RAM*.

SSADM Acronym for *structured systems analysis and design method*. As its name implies, it is a formalised method for designing computer systems. It has been adopted as a standard for public sector computing in the UK.
See also *structured analysis and design*.

SSI Abbreviation of *small scale integration*.

stack A mechanism for temporary storage of data in the course of processing, consisting

282

of a designated area of memory. New items of data are placed one by one at the top of the stack, which pushes items already in the stack down one place. Items are also taken one by one from the top of the stack, which cause remaining items in the stack to 'pop' up one place again – hence the longer names of 'push down stack' or 'push down, pop up stack'. A stack is a convenient way to queue data within a program because of the way many programs are constructed. To simplify them, programs are broken down into subroutines or procedures, called from a main routine. Subroutines may again call their own subroutines, and sometimes this continues down to several levels (or nests) of subroutine. Each subroutine pushes its temporary data on to the head of the stack, then pops it out again when it has finished. This leaves the stack exactly as it was left by the routine that called the subroutine, with no risk of one routine overwriting the temporary data of another.

See also *heap*.

stages of growth A planning method for information systems, developed out of a theory first put forward by Richard Nolan in the 1970s. Nolan identified four stages of growth in data processing, later extended to six, based on observable patterns of expenditure and organisational learning. stages of growth planning aimed to identify the stage that an organisation had reached, and hence what priorities it should set for expenditure and what controls it should apply.

Compare *business systems planning*; *critical success factors*.

standalone Of a computer system that operates independently of others, rather than connected to them via a network for direct exchange of information.

standard A specification of the way a piece of equipment should operate (or, often, of the way two pieces of equipment should interact – referred to as interface or communications standards), that is widely accepted. Standards in the information technology industry are of two main types:

(1) official or de jure standards, promulgated by official standards-making bodies such as ISO and CCITT;

(2) de facto standards, defined by leading manufacturers (and particularly IBM) and incorporated in their equipment, and subsequently adopted by other suppliers.

standard function A function supplied built into a programming language. A standard function does not need to be declared within a program but can be called just by using its name.

standard general markup language (SGML) A draft ISO standard for preparing text in machine-readable form. It is being used for an electronic version of the Oxford English Dictionary, intended to be made available on compact disk.

standard interface A standard way of connecting a peripheral to the processor of a

283

computer system. Each peripheral is plugged into the processor by means of a standard multipin connector which carries all the wiring necessary to exchange data and control signals. This means that a standard processor can be fitted with any number and type of peripherals (limited by the number of channels available), and can later be expanded with minimal hardware changes.

standby Held in reserve and ready to take over in the event of a failure.

star network See *star topology*.

star ring topology A topology, used for data communications networks, consisting of a central hub feeding a number of connected loops.

star topology A topology, used both for data and voice communication networks, in which all devices are attached via a central hub, such as a telephone exchange, a computer system or a controller.
    See also *active star*; *passive star*.

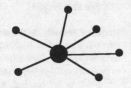

start bit In asynchronous transmission, a single bit transmitted immediately before the data bits, warning the receiving terminal that another character is on its way.

start of text (STX) A control character used to mark the start of the text of a message in a transmission block, normally following any address information.

start-stop transmission See *asynchronous transmission*.

stat mux Abbreviation of *statistical multiplexor*.

state transition diagram A representation of the expected behaviour of a computer-based system. It shows the states through which the system is expected to pass and the triggers which move it from one state to another.
    Compare *data flow diagram*.

statement An instruction to the computer originated by a programmer in a programming language. Statements are subsequently translated by a compiler or interpreter into the machine code instructions executed by the processor. In assembly languages, the statements correspond closely to the instructions executed by the processor. Conversely, in high-level languages statements correspond more closely to human methods of communication, whether verbal or mathematical, and each statement may generate a number of machine code instructions.

284

statement separator A symbol used to separate two statements in a program. In Pascal, for example, semi-colons are used as statement separators.

static binding Where the connection between the name of an operation and the program code to carry out the operation is established when the program is compiled, rather than at run time.

Compare *dynamic binding*.

static memory See *static RAM*.

static RAM (SRAM) A relatively expensive form of semiconductor random access memory that, unlike dynamic RAM, does not need to be refreshed periodically by the computer processor.

statistical multiplexor (stat mux) A multiplexor that obtains an increased effective bandwidth from the high-speed line on to which it is directing traffic. These devices use statistical techniques to allocate the bandwidth of the high-speed line, rather than allocating a channel continuously to each of the incoming low-speed lines.

status word A word in the memory of a computer system, set by the operating system and/or by peripherals, that contains a number of indicators and codes reflecting the current status of a particular system resource – a program, a peripheral, the system as a whole.

step-by-step operation See *single step operation*.

stepwise refinement A way of developing a program in small steps, used with block-structured programming languages such as Pascal. First the main program is written, but whenever a complexity or a problem is encountered it is given a name and passed over for the time being. Then each of those named problems is tackled in the same way. This process of successively decomposing the overall problem into subproblems is continued until the remaining subproblems can easily be coded. Each of those subproblems forms a separate 'block' – a function or procedure containing its own data. Also known as *evolutionary refinement*.

stiction A problem that sometimes occurs with hard disk drives where the read/write head sticks to the storage medium.

stop action video See *freeze frame video*.

stop bit(s) In asynchronous transmission, one or two bits transmitted immediately after the data bits making up each character.

storage compaction See *memory compaction*.

storage fragmentation See *memory fragmentation*.

storage location See *memory location*.

storage schema See *internal schema*.

storage tube display A method for creating the image on a visual display screen. Storage tube displays use a green phosphor on a dark green screen which is kept glowing by a low-level electrical charge.

store-and-forward switching See *message switching*.

stored procedure A pre-developed routine that other programs can bring into action when required. In a client-server arrangement, for example, stored procedures might be written to carry out complex or lengthy database changes and would be held on the server. Client programs would activate a stored procedure with a single call to the server, thus reducing the communication overhead.

stored program See *program*.

stored program control (SPC) A device controlled by a program rather than, say, by electromechanical logic. The term was coined to describe the first telephone exchanges that included a processor and a stored program to control switching, rather than electro-mechanical logic.

strategic data model See *corporate data model*.

streaming tape drive A cartridge tape drive using a particular technology that increases data storage capacity and transfer rate. Streaming tape drives are frequently used to back up hard disks attached to small computers.

string A series of characters treated as a unit, such as the characters forming someone's name or a warning message. The string is a data type in a number of programming languages, including BASIC and PASCAL.

striping drive A disk drive that writes each bit of a byte that is being recorded in parallel to eight different disks. Conventional disk drives, by contrast, write data serially to a single disk. The effect of striping is to allow an eightfold increase in transfer rate.

stroke refresh display A method for creating the image on a visual display screen. An electron beam scans the screen randomly, drawing lines and arbitrary shapes as required.

It must constantly retrace or refresh the displayed image for it to remain stable and visible.

structured analysis and design Methods for designing computer applications and programs using standard analysis processes and exercises. As for structured programming, the design is worked out progressively in more and more detail. A number of different techniques are used to achieve this, which can be separated into three main schools of thought:

(1) The data flow (or functional analysis) school uses models of the flow and transformation of data as the main thread of the design process. These are expressed in the form of (layered) data flow diagrams, process descriptions and a data dictionary. When a data flow model has been established describing the required logical behaviour of the application, technical options for implementing the application are analysed. The Yourdon method and SSADM (adopted as the standard for UK Government computing) are examples of this school.

(2) The Jackson method also sets out to model the behaviour of the application. The design is built up by defining processes which simulate the behaviour of their real world counterparts and which generate the required outputs. These are refined in a carefully developed sequence of design decisions, leading to the final implementation.

(3) The entity-relationship school focuses on the structure of the data that an application will deal with. These methods begin by analysing the characteristics of the entities (objects, events and concepts) affected by the application, and the relationships between them. These are believed to be more stable over time than the processes that handle the data. This analysis produces a logical structure for the data, known as a schema, from which is derived a schema for the data as it is to be stored physically.

structured design See *structured analysis and design*.

structured programming Programming techniques in which the program is described and then written in an agreed, systematic way. Usually only a limited range of control structures may be used. The program may also be broken down into modules – see modular programming. Structured programming is intended to reduce design errors and make programs easier for others apart from the author to understand and to maintain. The techniques can be applied with any programming language, but are explicitly supported by so-called block-structured languages such as ALGOL and Pascal.

Compare *modular programming*.

structured query language (SQL) A query language developed at IBM's San José Research Labs, based on the relational model. It was originally called SEQUEL (Structured English QUEry Language). It has since been adopted by a number of suppliers of relational database management systems. A standard definition of the language has been produced by ANSI, but nonetheless implementations of the language vary and many have proprietary extensions introduced by suppliers to suit their own products.

SQL is neither a simple enough language to be used directly by end-users to retrieve data from files, nor is it a complete programming language like COBOL or BASIC, since it only describes the organisation and retrieval of data. To create useable applications, it must either be embedded within an applications package or be used from another programming language that can include statements using SQL syntax.

Sample SQL queries
(computer output in bold, explanation in italics)

List all employees in department 9 still due for review (i.e. 90 days after hiring), in review date sequence

```
SELECT    Name, Hired, Hired + 90
FROM      Employee
WHERE     Hiredate + 90 > Sysdate
AND       Dept = 9
ORDER BY  Hiredate;
```

Name	Hired	Hired+90
Willett	11-Mar-91	9-Jun-91
McCall	21-Mar-91	19-Jun-91
Craig	11-Jun-91	9-Sep-91

List name, salary and department of all employees with above-average salaries

```
SELECT    Name, Salary, Dept
FROM      Employee
WHERE     Salary >
          (SELECT AVG (Salary)
           FROM Employee);
```

Name	Salary	Dept
Brown	15,000	12
Jones	23,750	12
Smith	19,600	9

---

structured walkthrough A technique used to check programs written using structured programming techniques. The programming group check the program by 'walking' through its logic to verify that it does what is intended by the program specification.

structured wiring Installation of components and cables to support the interconnection of telephones, computer and office equipment within a building or within a group of related buildings. This is usually done when buildings are constructed or modernised. Apart from the cables or other transmission media, structured wiring schemes also include *cross connect systems* (also known as *patch panels*) and connection points. It is more expensive initially than conventional wiring, but makes the subsequent job of managing communications systems easier and less costly.

Compare *block wiring*.

STX Abbreviation of *start of text*.

style sheet A coded description of the format of a paragraph, recorded by a word processor under a particular name or identifier. Once a style sheet has been defined – for example by specifying fonts and line spacing to be used for headings and text, sizes of margins, etc. – it can be transferred to a new paragraph or into a new document to save the labour of re-specifying all these details.

sub-schema See *external schema*.

subject area database A database containing data which is organised around particular business entities or subject areas, such as customer or quality control, rather than around individual applications, such as order processing or manufacturing scheduling. Many different applications programs may share data from a single subject area database.

subrange A variable declared as only being allowed to take on values in a subset of the full range available.

subroutine A subsidiary routine in a program that can be called repeatedly from another routine whenever it is needed, rather than the statements it contains being recoded in the calling routine on each such occasion.

Compare *in-line coding*.

subscript An index value used to access an element within an array, by specifying its relative position in the array.

subsystem A system that forms part of a larger system. Large computer systems, for example, may include several disk subsystems, each consisting of a disk controller connected to a number of disk drives.

super-user A non-specialist user, normally of personal computer applications, who has developed particular expertise and who helps other staff working with him to solve their computer-related problems.

supercomputer A computer system with a very powerful processor designed to handle mathematical calculations involving very large numbers of variables, such as in weather forecasting.

superconductivity The effect seen when electronic circuits are cooled to very low temperatures. This greatly speeds the flow of electrons and thus improves circuit performance.

supergroup A telephone cable capable of carrying 60 simultaneous calls.

supertwist A technology for liquid crystal display (LCD) screens, used on some portable computers. It has inferior performance – less brightness and a narrower viewing angle – but is cheaper than an alternative technology known as *active matrix*. Also known as *passive matrix*.

swapping The transfer of a page of program or data from memory to mass storage or vice versa.
See also *virtual memory*.

switched multimegabit data service An high-speed digital transmission service available in a number of countries. It is a connectionless service (i.e. messages are just 'posted' into the network) operating at speeds between 1.5 and 45 megabit/second.

switched network A communications network that transports information to a number of different destinations, depending on the address information supplied by the devices using it. Contrast with an *unswitched network*, such as a typical network of teleprocessing terminals, where information follows predefined routes between the terminals and the central computer system.

switching equipment Equipment that directs communications traffic on to alternative transmission lines, such as telephone exchanges for voice traffic and packet switching exchanges for data traffic.

SYLK format Abbreviation of *symbolic link format*.

symbol set See *character set*.

symbolic address The form an address takes in a source program, such as the line number (as in BASIC) or the identifier (as in most other languages) used for a statement, a procedure, or a data field. Symbolic addresses are translated into absolute addresses in memory when a program is compiled or interpreted.

symbolic language See *source program*.

symbolic link format (SYLK format) A format for exchanging information between spreadsheet applications on personal computers, originated by Microsoft.

symbolic name See *identifier*.

SYN Abbreviation of *sync character*.

sync character (SYN) A special character used in synchronous transmission techniques. One or more sync characters are sent immediately before each message, to align the clocks at the sending and receiving devices.

synchronous data link control (SDLC) A bit-synchronous data communications protocol, defined by IBM as part of systems network architecture (SNA) and supported by most of IBM's current products as well as by some of its competitors'.

synchronous digital hierarchy (SDH) A technology for multiplexing broadband digital signals such as those carrying video or speech. It can combine incoming traffic from a number of circuits operating at 2 megabytes/second or faster on to a single circuit running at up to 2.5 gigabytes/second. It overcomes many of the drawbacks of a predecessor technology, *plesiochronous digital hierarchy*, and is also far more flexible.

synchronous transmission A method of data transmission in which messages are sent as a continuous string of characters without gaps. Synchronisation of sending and receiving devices is established by sending special synchronisation (SYN) characters immediately before each message, to align the clocks within each device. Higher transmission speeds can be achieved with synchronous transmission than with asynchronous transmission – up to 9600 bit/s over the public telephone network. Synchronous transmission is widely used for teleprocessing.
See also *binary synchronous communications*; *bit-synchronous*.

syntax The set of rules describing what constitutes a grammatically correct program, written in a high-level language. Usually, the compiler will refuse to accept statements that infringe the syntax rules.

system A collection of interrelated objects interacting in order to meet certain defined objectives. According to systems theory, a system has three characteristics:
(1) Emergent properties, or in other words properties belonging to the whole system that are meaningless for its parts. For example, a car is a means of transport, but none of its components can be described in that way.
(2) Hierarchy: reality consists of layers of systems within systems. For example, an information system may consist of a number of computer-based and manual systems, the former in turn comprising several computer systems.
(3) Processes of communication and control that enable it to adapt and thus survive in a changing environment.
Every aspect of a working environment can be seen as a system – the physical environment, consisting of offices, buildings, etc; the way in which work is organised,

the operating procedures, the local habits and customs, the working relationships between people involved; and the tools and equipment used. Within the field of information technology, the term is applied at a number of different levels. Three uses are particularly widespread:

(1) In a broad sense, including people as well as any computer-based equipment, as in 'Effective (information) systems are crucial to business success';

(2) To mean computer-based equipment and associated procedures alone, as in 'The (payroll) system runs monthly';

(3) To mean the computer equipment alone, as in 'The (computer) system failed twice this morning'.

system building tools (SBT) Programs intended to help system analysts and programmers to develop applications programs. The term is normally used to mean programs that provide more help than conventional programming languages, such as COBOL or PL/1, and they can reduce both the effort and the time needed to develop applications programs in comparison with those, sometimes dramatically. This is usually achieved at the price of lower run-time efficiency, again sometimes dramatically so, and reduced flexibility. It may not be possible, for example, to design screen formats exactly as they are required. System building tools vary widely both in terms of the type of program that they can be used to build, and in terms of their power and sophistication. Some operate in conjunction with a database management system, and this class of tool is often referred to as a fourth generation language, or 4GL.

system development The process of designing, building and installing a computer-based system. The conventional system development process, used for most large and complex applications, consists of a number of more or less consecutive phases. These are:

(1) feasibility study or analysis of requirements, which determines what the system should aim to achieve;

(2) systems analysis and design, which determines how the system as a whole should be structured, including its data files, and how it should fit into existing procedures;

(3) programming and testing, which turns the results of the preceding phase into a suite of working programs;

(4) installation, to deliver the system to its users and bring it into operation.

During the subsequent life of the working system, it is likely to undergo a continuing process of enhancement and maintenance. This phase may or may not be regarded as part of the system development process.

The drawback of this conventional process is that it is very time-consuming, and requirements may change, perhaps drastically, while it is in progress. To avoid this danger, an iterative development process is sometimes adopted, using techniques such as *prototyping*, in which additional detail is added on each iteration. An abbreviated systems development process is often adopted for small systems; for systems developed by end-users rather than by a specialist information systems department; and where powerful system building tools are used. See also *development lifecycle*.

system documentation A general term meaning the various documents that are produced describing how a computer-based system has been constructed and how it operates. These documents are used both by people who maintain the software (the technical documentation) and by those who use and operate the system (the user documentation). The figure below illustrates the system documentation that might be produced for a large system, showing at what stages in the development process it would be produced. See individual entries for detail of the content of the various documents shown in the figure.

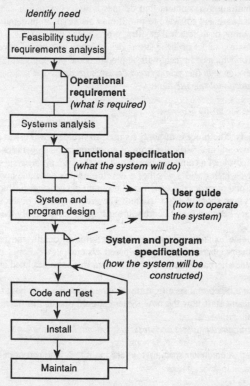

system generation A process used to initialise the operating software of large computer systems. It involves running various utility programs that build tables and routines into the operating system to reflect the chosen hardware configuration and software options.

system implementation The process of delivering to its intended users the programs that constitute a new computer-based system, and bringing into operation the new procedures required to use them.

system reset See *reset*.

system software See *operating software*.

system specification A document that defines how a computer-based system will be expressed in hardware and software terms. It may also explain in detail how the system will operate, in terms of the clerical or other procedures that it affects. The system specification is the main product of the systems design stage of systems development, and it forms the basis on which equipment will be procured and programs will be designed.

See also *development lifecycle*; *system documentation*.

Compare *functional specification*.

system utilities See *utility program*.

systems analysis The process of analysis used to decide how best to put a computer-based system to work in a particular area of an organisation's operations. In the normal development lifecycle of a computer-based system, the systems analysis stage follows the feasibility study, which decides whether a development is desirable and what its objectives should be, and precedes the system and program design stages, which are concerned with how the system should be constructed. The precise content of the stage varies, and also the boundary between it and neighbouring stages, but generally the systems analyst will:

(1) investigate existing systems and procedures in detail, and redesign them to accommodate the proposed new computer-based system;

(2) assess the alternative ways that the system might be designed and decide on the best option;

(3) produce a functional specification designed to make clear both to potential users and to development staff how the new system should operate, in functional rather than technical terms.

See also *structured analysis and design*.

systems analyst A computer specialist whose main task is to carry out systems analysis work.

systems applications architecture (SAA) IBM's blueprint for bringing together its diverse product lines, announced in 1987, and as such likely to be influential within the business computing world. In IBM's words, systems applications architecture (often abbreviated to SAA) is 'a collection of software interfaces, conventions and protocols, a framework for the development of consistent, portable applications in the company's three major computing environments – 370, System/3x, and PCs'. It is divided into four basic parts:

(1) A user access standard to ensure that applications programs operate in a consistent way as far as the user is concerned.

(2) A common programming interface so that applications can easily be transferred from one IBM computer to another.

(3) Common communications support, centred round IBM's existing SNA (systems network architecture) and DCA (document content architecture) specifications.

(4) Common utility applications such as office services and electronic mail.

systems approach A view of the application of technology to organisations that sees their activities and processes in terms of systems. Systems thinking has a similar general meaning, but refers in particular to a school of thought associated with Peter Checkland at Lancaster University that originated soft systems methodology.

systems design The process of translating a functional specification of a computer-based system (WHAT it should do) into a detailed design (HOW it will do it).

systems engineer A computer specialist who helps user organisations to put their plans for computer systems into effect. The term is used by some computer suppliers, and notably IBM, to describe the support staff they allocate to major customers to help them with this task.

systems house A business organisation that develops applications programs or other types of software for its customers and delivers and installs them along with the hardware on which they are to run. Contrast with *software house*, that generally supplies software alone.

systems inertia A term used to highlight the powerful influence that computer systems developed in the past have on what can be done today. This arises because new systems often have to use the same data files as existing systems, or fit into the same operating procedures. Because many old applications would be very costly to change and because information technology has advanced so rapidly, systems inertia has become a severe constraint on the information systems plans of many organisations.

systems integrator A business organisation that assembles and installs systems for its customers, pulling together the necessary hardware and software components from a number of different suppliers.

systems network architecture (SNA) IBM's blueprint for networks of computer systems and terminals, first announced in 1974. SNA was intended as a conceptual framework, specifying how terminals and computers would exchange information, and where the various network and processing functions would be located. It includes as a central component a bit-synchronous data link protocol known as SDLC. It has since been embodied in a large number of products, both from IBM and from its competitors.

It began as a framework for networks of data terminals surrounding a large central computer system, but has since been modified and extended to cater for distributed processing applications involving a number of computer systems.

Although the specifications of SNA are well known, it remains a proprietary scheme under IBM's control, incompatible with international interconnection standards such as the OSI standards being defined by the International Standards Organisation (ISO).

See also *systems applications architecture*.

Compare *open systems interconnection*.

systems programmer A specialist in computer operating systems (or, more frequently, one particular operating system). Systems programmers are needed in large computer installations to install and maintain the operating system and to deal with problems that arise from operating system malfunction.

systems programming See *systems programmer*.

systems thinking See *systems approach*.

systolic array A class of parallel processing architecture consisting of a very large number of processing elements (cells) interconnected into a regular structure. Each cell communicates with a fixed number of neighbouring cells and data items flow through them, synchronised by an external clock circuit. Wavefront array processors have a similar architecture but do not use an external clock circuit – data is transmitted as soon as it has been computed.

T

T1 carrier A high-speed transmission line available in the US, operating at a speed of 1.54 megabits/second.

tag (1) In general, a field added to a record to help to identify it or its status. For example, a tag may be added to a transaction record that goes through several processing stages to indicate what stage it has reached.

(2) Used specifically to mean a short record placed in the home block of a data file on direct access storage to indicate where an overflow record can be found. It usually consists of the key of the overflow record and the address of the block where it has been stored.

tagged image file format (TIFF) A format for files recording digital images in bit map form, originated by Aldus and Microsoft and supported by a wide range of scanner, image manipulation and desktop publishing software. The format records grey scale values at cell level, which means that images recorded in this format can be edited in detail on screen.

The format is designed to be flexible and extensible, and this has led to problems of incompatibility as equipment developers have interpreted parts of the standard in different ways. To address this problem and simplify programming, four TIFF classes have been defined: class B is for bilevel images where each pixel is either black or white; class G is for grey-scale images with up to 256 grey levels; class P is for palette colour images, where the palette defining the colours is included in the TIFF file; and class R is for full (24-bit) colour images with up to 16,77,216 shades.

Compare *encapsulated PostScript format*.

take on The capture of data to bring the files up to date prior to the introduction of a new computer-based system.

talk mode The state of a modem when it is not receptive to data signals, but instead allows signals to go straight through to the telephone handset.

Compare *data mode*.

tape A common form of mass storage, using magnetic tape similar to that used in domestic cassette recorders as the storage medium. For larger systems, 1/2 inch wide tape is the standard, held in 2400 foot reels or cartridges. Data is recorded in seven channels which run along the tape. Six of those channels record data bits, forming six-bit characters across the width of the tape, and the seventh is a parity bit. The capacity of a single reel varies depending on the packing density, and typically is several million characters. Transfer rates range up to several hundred characters per second.

Data is written on the tape as continuous sequences of six-bit characters, known as blocks. Blocks are written to or read from the tape in a single operation, and there is a gap between each block (the interblock gap) to allow the tape to slow down and speed up between read or write operations. Each block may be divided into logical records which are passed one by one to applications programs by operating software. Records are blocked in this way to economise on space on the tape and to increase processing speeds, since otherwise interblock gaps would take up more space than the data itself.

Smaller cassette and cartridge tape drives are used on home computers and to back up hard disks on small computers.

tape deck See *tape drive*.

tape drive A device for reading and writing magnetic tapes used for mass storage of data. It consists of a transport mechanism on which a tape reel, cartridge or cassette can be mounted, and read/write heads for transferring data to and from the tape.

tape library A storage area for magnetic tapes after or awaiting processing, or the tapes themselves.

tape mark A special single-character block on a magnetic tape, indicating that the block following is a qualifier block, rather than a block of data.

tape serial number A number uniquely identifying a reel of tape at a particular installation, held in the header label and usually written on the outside of the reel or cartridge. The tape serial number is assigned to a new tape and remains unchanged throughout the life of the tape, whereas the file name and other identifying information may change. It is used in the tape library and by computer operators to identify the tapes required for processing.

tape streamer See *streaming tape drive*.

tape transport See *tape drive*.

tape unit See *tape drive*.

tariff The charging rates and regulations applied to subscribers to a communications service by the operator of the service. Tariffs may have any combination of three components:

(1) A connection charge levied once when the subscriber first registers for a particular service.

(2) A standing charge, levied for each billing period (typically a month or a quarter) during which the subscriber uses the service.

(3) A usage charge, based on how heavily the service is used. This may be based on the number, duration, or type of calls made; and the volume of information transferred, the distance it covers, or the time zone in which it occurs.

TCP/IP Abbreviation of transmission control protocol/internet protocol. These are a set of standard protocols for communications networks, originally defined by the US Department of Defense for military use. They were defined as an interim measure, pending the completion of work on defining open systems interconnection (OSI) standards.These protocols are in widespread use within the US defence community and have also built up a following in the business computing world.

TCP/IP is based on connectionless working – in other words messages are consigned to the network like letters into the mail system – with TCP forming the transport layer (level 4) and IP the network layer (3). The set also includes higher level protocols for file transfer (file transfer protocol – FTP), electronic messaging (simple messaging protocol – SMP) and network management (simple network management protocol – SNMP). As their names suggest, these protocols are simpler and easier to implement than the equivalents defined by bodies such as CCITT and ISO, but they are also more limited in scope.

TDI Abbreviation of *trade data interchange*.

TDM Abbreviation of *time division multiplex*.

tear-off menu A refinement of the pull-down menus that are part of the graphical user interface found on personal computers such as the Apple Macintosh. As well as operating in the normal 'pull-down' way, these menus can be 'torn off' by 'dragging' the menu down the screen. The menu then detaches itself from the menu title at the top of the screen and can be placed anywhere on the screen, where it remains visible until removed. This means that menu options can be selected with a single press of the mouse button.

technical and office protocol (TOP) A set of supplier-independent protocols for data communication within the technical and administrative offices of manufacturing companies, first proposed by Boeing. The TOP protocol initiative followed General Motors' attempt to establish support for the MAP (manufacturing automation protocol) protocols which it had defined. Like MAP, TOP protocols are a subset (otherwise known as a profile) of the open systems interconnection (OSI) standards.

telebanking Applications or equipment that enable banking transactions to be carried out automatically remote from the bank's processing centre, such as from home or via automatic teller machines (ATMs).

Telecom Gold British Telecom's public electronic mail service.

telecommunications The transmission of information – data, text, image, speech, video – by means of electromagnetic signals. Information may be carried over short or long distances, using a variety of physical transmission media such as cables or optical fibres, and/or via radio or satellite circuits.

telecommunications network See *communications network*.

telecommunications regulation The rules governing the use of telecommunications services. These specify, for example, which carriers may supply which services; what equipment may be attached to public data networks; what traffic may be carried on telecommunications services leased from the carriers for private use. Regulation varies from country to country depending on policy for national services and political attitudes to monopoly provision. In the UK, enforcement of regulation lies in the hands of a Government agency called Oftel, and there is an equivalent body in the US called the Federal Communications Commission (FCC).

telecommuting Carrying out office work in the home rather than in the office, using information technology equipment and a telecommunications link to maintain contact. Also known as *homeworking* and, occasionally, *networking*.

teleconferencing Conducting a conference between a number of people at different locations, using telecommunications and other supporting equipment to convey information between them. Generally, teleconferencing implies that all the participants are present at the same time, and the links between them may be solely by telephone (audio conferencing) or include moving pictures as well (video conferencing), with a variety of aids for conveying graphics. But there is a related technology called computer conferencing where participants may be apart in time as well as space, in which a shared computer mediates the exchange of messages between them.

telematics See *information technology*.

telemetry The recording of measurements, such as electricity meter readings, and the automatic transmission of that information to a remote collection point, such as a computer system.

telemonitoring The automatic collection of data relating to the performance of equipment, and the transmission of that information to a remote collection point, where staff responsible for repair and maintenance of the equipment are located.

teleport An installation that provides access to a range of powerful telecommunications services, such as satellite transmission, videoconferencing, and so on. Some large cities have installed teleports to encourage businesses to site their offices locally. These offices and public buildings such as conference centres can be attached directly to the teleport by means of high-capacity transmission lines.

teleprocessing A form of processing in which transactions are entered at a number of remote data terminals and are processed in a centralised computer system. The system checks the details of transactions as they are entered so that errors can be corrected immediately.

A conventional computer is designed essentially to do one thing at a time, so, to achieve a high throughput of transactions, a software component called a *teleprocessing monitor* is used to allow applications programs to process a number of transactions concurrently. To improve processing efficiency, software techniques such as *multitasking* and *multithreading* are used. Multitasking permits shared access to disk files, while multithreading enables many like transactions to share one copy of the applications software without interfering with one another.

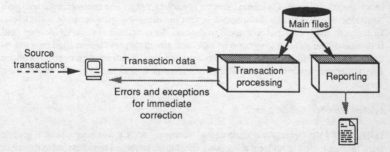

teleprocessing monitor A software package designed to support the development and operation of teleprocessing applications. These control large numbers of terminals connected via communications links to a shared computer system running the applications programs. The teleprocessing monitor serves as an intermediary between applications programs and the operating system, and between applications programs and the terminals. For example, it monitors the arrival of messages from terminals and makes sure that each is passed to the appropriate applications program.

teleshopping Shopping without leaving the home, placing orders by electronic means.

301

Customers have access to some kind of electronic catalogue that they can bring up on a screen via the telephone network. They then place orders via keyboard or keypad which are transmitted automatically to the supplier's warehouse. Goods are then delivered by van or mail.

telesoftware Software that is delivered over a telecommunications or broadcasting network direct into the user's computer, rather than being collected or mailed in physical form, such as on a floppy disk. Telesoftware can be obtained from many bulletin board systems, and is also broadcast by the BBC via teletext.

Teletel The trade name for the public videotex service in France.
 See also *Minitel*.

teletex An enhanced form of telex service, introduced by a number of European telecommunications authorities recently. Teletex terminals are more like word processors than conventional telex terminals, supporting upper and lower case text, and able to format and edit text freely. The teletex character set (the codes representing different characters) is a superset of ASCII. Not to be confused with *teletext*.

teletext A generic term for one-way broadcast information services which use an adapted television set to display a limited number of frames, consisting of text and simple block graphics. The BBC's teletext service is called Ceefax, and commercial television's equivalent is called Oracle. Information is sent out using the spare capacity in the television signal, and is captured and stored in the decoder contained in the television set until it is required for display. Contrast with *videotex*, which displays information on a television screen in a similar form, but is two-way and carries information over a telephone line rather than over the air. Should not be confused with *teletex* – the 'super-telex' service.

teletext decoder See *decoder*.

teletype (TTY) Originally a trade name, owned by AT&T, but now almost a generic term for dumb (that is, no local processing capability) keyboard terminals. Also describes the asynchronous transmission protocol, using ASCII character code, used by many such terminals and also by personal computers.
 See also *glass teletype*.

teletype protocol See *contention*.

teletypewriter See *teletype*.

telewriting Equipment which enables a drawing or writing being made at one location to be seen at another simultaneously.

template An empty document or file that has previously been set up for a particular purpose, so that the user only has to 'fill in the blanks' to produce the required result.

10Base2 See *thin Ethernet*.

10Base5 See *thick Ethernet*.

10BaseT A standard wiring scheme for Ethernet local area networks, ratified by the IEEE in late 1990. It comprises unshielded twisted pair (UTP) copper cables, running in a star topology from a concentrator to data outlets sited at regular intervals around the skirting or across the floor.
　　See also *structured wiring*.

tera- A prefix meaning millions of millions of (x 10^{12}).

terminal (1) A general term meaning any device used by people to gain access to or enter data into computer systems, including both keyboard and visual display devices, and devices such as remote batch terminals that are driven by punched cards. Terminals may be installed either distant from the computer system and connected by a communications link (known as a remote terminal), or close to it (a local terminal).
　　(2) Also used in a broader sense by telecommunications companies to mean any user device that may be attached to a network, including both computer systems and (in the narrow sense given above) their terminals, as well as telephones, facsimile transceivers, etc.

test data Data submitted to a program under development, designed to establish that it works correctly. Test data is designed by the programmer, or by a programming team leader or systems analyst, to present a program with all the conditions it is likely to meet.

test harness A program written to fit around a program routine or a module that is under development, so that it can be tested without having to use the whole program in which it is to appear.
　　See also *modular programming*.

testbed See *test harness*.

text (1) The information content of a message sent over a transmission link, as opposed to those elements of it that are there to facilitate transmission (the *envelope*).
　　(2) Information consisting of coded characters that can be printed on paper or shown on a display, and can then be read and understood. Contrast with *data*, that may not be meaningful when displayed, and with *image*, that does not consist of coded characters.

text editor See *editor*.

text file A data file whose records consist entirely of characters, excluding, for example, binary numbers.

text only format A format for information representing a document that includes only the content of the text. This means that, when it is transferred to a different computer or a different word processing package, formatting effects such as, for example, the size of margins, the position of tabs, or boldface and underline within the text are lost.
Compare *revisable text format*.

text retrieval A form of information retrieval where the files that are searched consist of free form text, such as documents, reports or abstracts, rather than structured data found, for example, in a spreadsheet or a personnel file. They may contain millions of entries (i.e. records), held either as coded text or as images of documents. The user either uses specified keywords (keyword retrieval) or any combination of words contained in the text (full text retrieval) as search criteria to identify the records required. Normally, the system reports how many records match the search criteria, and the user continues refining the criteria until a small enough number of records has been identified for these to be displayed one by one on the screen or be printed out on paper.

THEN statement See *IF statement*.

theorem proving See *predicate calculus*.

thermal printer A printer that uses heat-sensitive paper. The print head consists of an array of pixel-sized heater styli – either thick film resistors or semiconductors. These heat the paper so that it turns from light to dark. Thermal printers are light, reliable and cheap, but the paper is relatively expensive.

thermionic valve A component used to build the logic circuits of the earliest (known as first generation) computers. It consists of a number of metal elements surrounding a small heater, enclosed in a glass envelope. Thermionic valves were made obsolete by the invention of the transistor and subsequently of integrated circuits.

thick Ethernet The original version of Ethernet – a type of local area network. Since its arrival new versions of the network have been introduced, one of which uses thinner and cheaper coaxial cable, and for this reason the original is now generally referred to as thick Ethernet, or alternatively by the name of the relevant IEEE standard *10Base5*. Thick Ethernet networks can be laid over distances up to 500 metres.

thin Ethernet A version of Ethernet – a type of local area network – using thinner and cheaper coaxial cable than was used originally for Ethernet networks (now known, of course, as *thick Ethernet*). In consequence it is capable of operating over shorter distances than thick Ethernet: up to 200 metres compared with up to 500. Also known as

10Base2, which is the name of the relevant IEEE standard, and *cheapernet*. Compare *10BaseT*.

thin film memory See *bubble memory*.

third generation language See *high-level language*.

third normal form (3NF) The form which data assumes after undergoing the process of normalisation. It is a simple and consistent two-dimensional format in which:

(1) repeating data items (such as lines in an order) are included in separate tables, rather than with the data item to which they 'belong';

(2) all rows in a table depend on the entire key of that table, and do not depend on any other data item.

See also *relational model*.

third-party maintenance Maintenance services for information technology hardware provided by an organisation other than the hardware manufacturer.

third party reseller See *value added reseller*.

third-party traffic Communications traffic which neither originates nor terminates on a particular network.

See also *private network*.

32-bit Used mainly of a microprocessor or a bus (the data 'highway' within a small computer) to specify that it can handle 32 bits of data at once, as can the microprocessors and buses in today's high-performance personal computers. Together with speed of operation (expressed in MHz) this is an important indicator of performance.

3270 protocol A protocol defined by IBM for communication with its own 3270 family of clustered visual display terminals. It is now widely used for transaction processing applications. Expansion cards are also available for many personal computers which enable them to retrieve data from central computer files by emulating this protocol. The original version was binary synchronous (3270 BSC) and a later – bit-synchronous – version was defined to fit within IBM's systems network architecture (3270 SNA).

thrashing A condition of a computer system with virtual memory, in which it spends most or all of its time administering the virtual memory rather than in processing useful work.

thread One instance of a transaction.

See also *multithreading*.

305

3NF Abbreviation of *third normal form.*

throughput A measure of the amount of work performed by a computer system in a given period of time, such as transactions or jobs per hour.

thumbnail A representation in a reduced form of a page that is being prepared on a desktop publishing system.

tie line See *leased line.*

TIFF Abbreviation of *tagged image file format.*

tile To arrange windows on a visual display screen so that they do not overlap.

time division multiplex (TDM) Share a transmission channel by dividing it up into very short time slots, which are allocated in turn for a further chunk of the messages waiting to use it. This is like several lanes of traffic merging in an orderly fashion into a single-lane high-speed tunnel.
 Compare *frequency division multiplex.*

time-of-day clock A device within a computer system which registers the time of day, and sometimes the calendar date. It is set by the computer operator or user initially and whenever needed thereafter (such as when the system is switched off completely for repair). The time-of-day clock can be read by programs, via the operating system, for example to attach a timestamp to events that are logged or to include today's date in a letter.
 Compare *real-time clock.*

time out Wait for a predetermined time for an event to take place and, if it does not, initiate some kind of remedial action. For example, an operating system may apply a time out between the calls it receives from programs running under its control. If the time out expires, it can conclude that the program is in a loop or locked in a deadly embrace with another program, and will force the program to terminate. Devices sending messages across a transmission link also apply a time out on acknowledgements so that they can detect transmission failures. When the time-out expires, they retransmit the message, assuming that it has been lost in transmission.

time slicing A method used by computers to share processor time between programs. Processor time is divided up into fixed time slots that are allocated, either in strict succession or on a priority basis, to programs running within the computer.

timecode A signal recorded digitally on to videotape to identify each frame. In SMPTE format, for example, it takes the form hours:minutes:seconds:frames, as in 11:39:15:05.

timesharing A computing service supplied from a central computer to users at remote terminals. Users connect to the service by dialling in, either over the telephone network or over a data network, then enter commands and data to run programs, with results returned directly to their terminals. It is called timesharing because users share the processing time (and other resources) of the central system without being aware of other users connected at the same time. It has been widely used for scientific and technical computing, but is now being displaced by personal computers installed locally. Also known as *multi-access computing*.

Compare *teleprocessing*.

timesharing bureau See *service bureau*.

timestamp A data field containing the time of day and perhaps the calendar date, indicating when the event or transaction with which it is associated took place.

toggle A two-way option, such as a choice in a menu or a button on the screen. Whenever the user selects the option, it 'toggles' to show the alternative.

token bus See *token passing*.

token passing A method of controlling the traffic sharing a single channel of a local area network. The token, a special bit pattern, circulates continuously on the network. When a device attached to the network wishes to send a message, it waits until the token arrives, and can then send a block of information. After doing so, it must immediately release the token again so that other devices can take their turn. Token passing is used both on ring and bus networks, known as token ring and token bus respectively (and defined by ISO standards 8802/5 and 8802/4).

Compare *carrier sense multiple access*.

token ring A method of controlling traffic on local area networks – see *token passing*. Token ring is of special significance in that it is embodied in IBM's local area network product.

See also *Cambridge Ring*.

toll line See *trunk line*.

tool A component (usually software) that automates one of the activities within a method. Certain types of system development tool, for example, help programmers to produce flowcharts illustrating their programs.

TOP Abbreviation of *technical and office protocol*.

top-down Beginning with the whole, then progressively breaking this down into smaller units to eventually define in full detail. Computer-based systems are often designed in this manner. The reverse approach is known as *bottom-up*.

See also *functional decomposition*.

top-down programming See *structured programming*.

topology Of a network, the arrangement of the cables or other transmission media used to interconnect the devices using it. For teleprocessing networks centred on a computer system, star topology has traditionally been used. More recently, local area networks have been introduced which use more flexible topologies such as ring, bus and tree.

touch screen See *touch-sensitive screen*.

touch-sensitive screen A visual display screen that can detect when the user touches specified areas of the screen. These areas have sensors (either infrared or capacitance-sensitive) superimposed on the display. Menu options are displayed in the sensitive areas, which the user can touch to select.

touchtone dialling A form of dialling over telephone networks where the digits in the number are identified by tones of differing pitch. Touchtone dialling originated with pushbutton telephones and is also known as *pushbutton dialling*. Touchtone telephones can be used to enter small amounts of data such as in home banking applications.

See also *audiotex*.

Compare *pulse code dialling*.

Towers of Hanoi See *recursive*.

TP Abbreviation of *transaction processing*.

TP monitor See *teleprocessing monitor*.

TP1 A benchmark which is a cutdown version of Debit/Credit, allowing performance to be measured independently of the communications-related components, such as the teleprocessing monitor. Transactions are generated by a special generator program running within the computer system under test. Performance figures based on TP1 are widely quoted by suppliers of database management systems.

TPC Abbreviation of *Transaction Processing Performance Council*.

trace. See *program trace*.

track A longitudinal (in the case of tape) or radial (in the case of disk) channel for recording data.

trackball See *tracker ball*.

tracker ball A device used to control and enter information into a graphics workstation or personal computer. It consists of a ball set on top of a small case. The ball is moved around by hand, and those movements are reflected in the movements of the cursor on the display screen.
Compare *mouse*.

tracking Adjusting letter and word spacing uniformly over a range of text.
Compare *kerning*.

TRADACOMS Standards for the format of electronic messages used for stock ordering, invoicing and similar business transactions, abbreviation of *TRAding DAta COMmunicationS*. These standards were defined by the Article Numbering Association, an association of UK business organisations involved in retail distribution.

trade data interchange (TDI) A term formerly used for electronic data interchange, perhaps more accurately representing what that is.

traffic The flow of messages across a communications network or along a transmission line. Traffic on voice (telephone) networks is normally measured in erlangs during a busy hour. Traffic on data networks is normally measured in bits transferred in a given time period.

traffic analysis Formal and systematic measurement of the patterns of traffic on a network, followed by mathematical analysis of the results. Traffic analysis is used by operators of communications networks to identify weaknesses in network performance and to plan improvements.

traffic statistics Figures representing the pattern of traffic on a communications network, such as number and duration of calls, error rates, volumes of data transferred, and so on. May be used for network planning or for customer billing.

transaction All the processing activities – updating of files, returning an acknowledgement, etc. – that must be completed within a computer system to record an external transaction, such as the placing of an order by a customer or a purchase in a shop. Where the data for a transaction is entered via terminals (commonly called teleprocessing), a transaction might consist of a number of exchanges of messages between the terminal and the computer system containing the applications programs. Thus, in an order processing application, a transaction might consist of entering the customer's identity, followed one by one by all the items in the order.
Also used to mean the message or messages that are generated and sent to the computer system to record the transaction.

transaction file A data file containing information recording recent events, such as orders that have just been placed or recent changes to customers' addresses.

Compare *master file*.

transaction journal A file created by a teleprocessing system to record, in time sequence, all transactions and other events that have affected the contents of the data files. In the event of a machine breakdown or a software failure that damages the files, these can be recovered from an earlier file dump, then the transaction journal can be used to bring the data files up-to-date prior to restart.

transaction monitor See *teleprocessing monitor*.

transaction processing (TP) Data processing where transactions are processed as they arise, rather than being assembled into a batch that is processed later (batch processing). Generally, transaction processing systems deal with a number of streams of transactions at once, arriving from different data terminals. Throughput is measured in transactions per second, determined by dividing the concurrency – the number of transactions being processed within the system at a time – by the average occupancy of a transaction – the time taken from arrival in the system to completion of processing and despatch of a reply. See also *teleprocessing*, one of the most widely-used methods of processing transactions.

Compare *batch processing*; *sequential processing*.

Transaction Processing Performance Council (TPC) A body formed by a number of leading computer suppliers to create a standard benchmark for transaction processing systems. They have endorsed the Debit/Credit benchmark, and a cutdown version known as TP1.

transaction telephone A telephone equipped with a magnetic stripe reader. Credit or charge cards are 'wiped' through the slot past the reader, and the account number is read automatically and transmitted down the telephone line. Additional details, such as the amount of the transaction, can be keyed in using the telephone push buttons.

transborder data flow The transmission of data across national borders. Problems with transborder data flow are often legislative rather than technical because of, for example, differing provisions for data protection and differing regimes for telecommunications regulation from country to country.

transfer rate See *data transfer rate*.

transistor A device invented by John Bardeen, Walter Brittain and William Shockley of the Bell Telephone Laboratories in New Jersey in 1947 that replaced the thermionic valves used in the original computers. It is produced by introducing impurities, known as dopants, into semiconductor material such as silicon. Pure crystalline silicon is a good electrical conductor, but when 'doped' with these impurities shows either an excess (n-type, for negative) or a deficiency (p-type, for positive) of electrons when a voltage is applied. If a sandwich (n-p-n or p-n-p) of these materials is created, the current flow between two of the elements can be regulated by a voltage applied to the third. Originally, transistors were manufactured in a small metal envelope, to be wired separately into electrical circuits. They are now incorporated directly into integrated circuits.

transistor-transistor logic (TTL) A method of designing and constructing the logic in integrated circuits. Transistor-transistor logic is cheap and flexible, but has the disadvantages of high power dissipation and limited speed.

transit time See *propagation delay*.

transmission block A sequence of data bits or characters transmitted as a unit. Error detection and correction procedures are applied at this level, so the maximum length of transmission blocks is normally set based on the likelihood of errors and the time needed for retransmission when errors are detected. Long data files and messages are therefore often split down into blocks for transmission, and are reassembled on receipt. In packet-switching, transmission blocks are referred to as packets.

transmission channel A path between two devices to carry information between them.

transmission line Telecommunications equipment directly connecting terminals, computer systems or communications control devices so that messages or other forms of information can be transferred between them. A single transmission line can be divided into a number of channels in either direction so that a number of exchanges of information can be taking place simultaneously (see *multiplexor*, *frequency division multiplexing* and *time division multiplexing*). Transmission lines vary in physical characteristics such as speed and reliability, depending on the type of transmission medium and supporting equipment that are used.

transmission link See *communications link*.

transparent Not perceived by the initiator of an operation. In other words, the hardware and software takes care of the details without involving the initiator.

transport layer The fourth protocol layer of the OSI reference model, defined by ISO. The transport layer ensures that there is a reliable path for data all the way between the sending and receiving devices.
See also *application layer*; *data link layer*; *network layer*; *physical layer*; *presentation layer*; *session layer*.

transport level See *transport layer*.

311

transport network In general, the part of a communications network that transports information, as opposed to the access network that enables users to connect to it and request its services.

Used more specifically to refer to a type of network service that carries customers data in specified formats and using specified protocols.

transputer A complete computer on a single integrated circuit, especially adapted to operate with others in a parallel processing arrangement. The transputer was developed by Inmos, a microelectronics company originally set up by the UK Government but now in the private sector.

trap Prevent an instruction from executing and instead enforce a branch to some other routine. Traps are mainly used by operating systems to prevent illegal instructions from causing damage to programs or data in memory at the time or to transfer execution to a routine in read only memory (ROM).

tree structure A hierarchical structure for a data file or database in which records contain links to one or more records lower in the structure. Users looking for information start at the top of a hierarchy and work their way down by following the links until they find the record containing the information they want. Videotex databases are constructed in this way. Records in the higher levels of the hierarchy that contain nothing but routing information are known as index or routing pages.

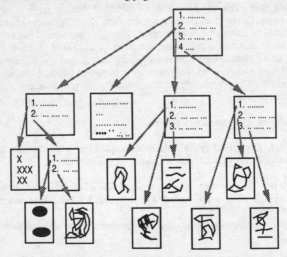

tree topology A topology, used both for local area networks and for teleprocessing networks, in which transmission links to the devices split like the branches of a tree. Poll/select protocols are normally used to control the traffic on teleprocessing networks, and carrier sense multiple access (CSMA) protocols on local area networks.

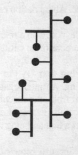

trigger In the context of database management systems, an event which automatically triggers off other transactions. A programmer might define the despatch of the final item in an order, for example, as a trigger to generate the invoice.

true BASIC An extended version of the original BASIC programming language (the one that is widely available on personal computers). In comparison, it is a structured language with features to support the writing of large and complex programs, and has extended capabilities in terms of graphics (including animation), matrix operations and sound generation.

trueType A typeface technology jointly developed by Apple and Microsoft, to be used on Apple's Macintosh and IBM's PS/2 range of personal computers. TrueType fonts are intended to be used via the PostScript page description language. Like PostScript fonts, trueType fonts are outline fonts – in other words the shape of each letter is described by means of a series of bezier curves. Unlike PostScript fonts, they also include what are known as hinting instructions, which ensure that outlines can be scaled to virtually any size and still maintain their shape, and also that they look well both on low-resolution devices such as display screens and on high-resolution printers or typesetters.

trunk circuit See *trunk line*.

trunk exchange See *exchange*.

trunk line A long-distance transmission line, normally carrying a number of multiplexed traffic streams.
　　Compare *local line*.

trunk network See *backbone network*.

TTL Abbreviation of *transistor-transistor logic*.

TTY Abbreviation of *teletype*.

tuple A row in a relational database file, in other words all the data fields describing one occurence of an entity. This is analogous to a record in a conventional file.

Turing test A test for artificial intelligence, proposed by the mathematician Alan Turing. An interrogator is seated at a terminal that is connected to, but physically separated from, the machine to be tested, for which a human being may be acting as substitute. If the interrogator cannot tell whether the answers to his or her questions are coming from a human being or from a machine, Turing argued, then the machine can be said to think.
　　See also *knowledge processing*.

turnaround document A document, printed by computer, that is used to initiate a clerical operation which adds data to it, and is then returned to the computer to be re-input to update the data files. For example, documents might be produced with names and addresses of public utility customers whose meters were to be read. The meter reader would then mark the readings on the documents, which would be returned and be read back by the computer to produce gas bills. Mark sensing and optical character recognition technologies are often used to read such documents.

turnaround time (1) The time taken by a modem attached to a half-duplex transmission line to reverse the direction in which information is transmitted along the line, in other words to switch from send to receive or vice versa.
　　(2) The time a teleprocessing system takes to process a transaction received from a terminal, from its arrival at the system to the despatch of the final reply message.
　　Compare *response time*.

turnkey supplier A business organisation that takes complete responsibility for developing and installing a computer system, complete with applications, as specified by a customer, including if required training customer staff in its use.

24-bit Most often used to specify how many bits are used to achieve a particular effect, as in '24-bit colour'. Since 24 bits can carry 2^{24}, i.e. over 16 million, combinations, 24-bit colour display screens are capable of displaying that number of different colours. This produces a picture quality very close to that of high-quality photographs.

2780 protocol A protocol defined by IBM for its own product, a remote job entry terminal coded the 2780, and widely used for remote job entry and file transfer. The original version was binary synchronous (called 2780 BSC) and a later – bit-synchronous – version was defined to fit within IBM's systems network architecture (2780 SNA).

twisted pair The simplest and most common transmission medium used for telephone wiring, both inside buildings and out to the local exchange. It consists of two insulated wires, for the signal and for an earth return, in a plastic sheath. The wires are twisted within the sheath to reduce the risk of crosstalk when they are carried in the same cable or duct as other pairs.

two-phase commit A procedure enforced by database management software to ensure that a transaction that updates more than one database file does either all or none of the updates. If the transaction fails to complete the final phase of processing, such as because a remote computer holding one of the files is unavailable, all the updates it has made so far are reversed automatically.

two-wire circuit A transmission line consisting of a single pair of wires, most commonly used within buildings and over short distances to local exchanges. Four-wire circuits are used over longer distances and for multiplexed lines.

typeahead Where an applications program records and stores what is typed on a computer keyboard, even when it is unable to change the display quickly enough to keep up. As soon as the typist pauses, the program catches up again so that the display accurately reflects what has been typed.

typeface A style of type design. Typefaces usually have a distinctive name, such as Times, Helvetica, Avant Garde, and so on.

typestyle A typeface in a given font size and a given style – roman, italic, bold, etc.

U

UCSD Pascal A modified version of Pascal developed at the University of California in San Diego and widely available on personal computers.

ULA Abbreviation of *uncommitted logic array*.

unattended operation A mode of operation of a computer system or a terminal that does not require a human being to be present.

See also *darkroom computing*.

unbundle To sell individually components of a computer system previously available only at a single inclusive (i.e. bundled) price.

uncertainty Ways of handling imprecise information in expert systems and decision support systems.

See also *combined fact-rule uncertainty*; *fuzzy logic*.

uncommitted logic array (ULA) A type of integrated circuit manufactured initially without connections between a number of its logic elements. In a second processing stage, these connections are built into the circuit. This is a relatively inexpensive way of making circuits of which only a few copies are required. The circuits will, however, be less efficient than circuits that are manufactured in fully integrated form – application specific integrated circuits.

unconditional branch See *GO TO statement*.

unicode A character set being developed by a consortium of major computer suppliers intended to make multilingual software easier to produce. Today's widely-used character sets, such as ASCII, use only 8 bits to describe each character. Unicode uses 16 bits, which means that special characters can be given an unique code regardless of language.

unit A single physical device. Peripheral and terminal subsystems often consist of a number of units attached to a single controller.

unit record device A general term meaning peripherals such as card readers that hold each unit of information as a separate physical record.

Universal Product Code See *bar code*.

Unix A portable operating system, able to operate on a wide range of computer systems from mainframes down to personal computers, and particularly well suited to engineering applications and to program development. It was originated by staff at Bell Laboratories for their own use with minicomputers, and has since been promoted by AT&T and by a number of major computer suppliers as a manufacturer-independent standard for operating systems. The strength of Unix as a portable operating system is that a 'shell' can be written for it to accommodate the particular human/computer interaction and data management procedures used on a particular machine. Unfortunately, this means that versions of Unix on different machines may both appear different, and be functionally different, and a number of alternative 'standards' for Unix are competing for acceptance.
  See also *POSIX*.

unshielded twisted pair A type of copper cable used for in-building data networks. High-grade unshielded twisted pair wiring is capable of supporting data transfer rates up to 100 megabits per second.

unswitched network A network where information follows predefined routes between the terminals. Many teleprocessing networks take this form.
  Compare *switched network*.

up time The time that a computer system is switched on and available for service.
  Compare *down time*.

update Change the contents of a data file, a program or a variable to reflect the latest required values.

upgrade Of hardware, install additional capacity, such as more memory, a faster processor or more disk drives. Of software such as an operating system or an applications package, acquire a new or improved version.

upload Transfer (a program or a file) from one system or device in a network, to be processed by another higher in the hierarchy of control.
  Compare *download*.

upwards compatible Useable with a later or more powerful version of a product. The term is most often used to describe products that have been superseded (in the supplier's terms) by a new one. If the superseded product is described as upwards compatible with the new version, it means that procedures and programs developed to operate with that earlier version will continue to operate properly, without modification, with the new version, but not vice versa.

Compare *downwards compatible*.

usage charge A charge for a communications service, based on how heavily the service is used. This may be based on the number, duration, or type of calls made; and the amount transferred, the distance it covers, or the time zone in which transfer takes place.

user Originally used within the computer industry to mean the purchaser of computer equipment, which was usually the data processing department of an organisation. Another term was needed for those people in the organisation who were the real users, in the sense that the computer system ran applications programs, generated reports, and so on, to help them do their jobs. 'End-user' has been adopted with that meaning, although 'user' is sometimes used in this sense also.

user coding See *own coding*.

user data The part of a message (or packet or transmission block) that contains information of interest to the application in the receiving device, rather than the control information or envelope that serves to regulate the transfer.

user-friendly Describes computer systems and applications that non-specialist users find easy to use. Sometimes this derives from simplicity – see for example *videotex* – sometimes from sophisticated interface mechanisms – see *graphical user interface*.

user group A voluntary association of the users of a particular hardware or software supplier's product. User groups meet regularly to exchange experience with the product, to discuss problems and, where appropriate, to put pressure on the supplier to improve the product or to change commercial policy.

user-view schema See *external schema*.

utility program A program, usually supplied by the manufacturer of a computer system or a software package, that performs a particular housekeeping or maintenance tasks. Utility programs are usually provided by a computer supplier, for example, to make security copies of mass storage files; to convert to a useable format files created with another manufacturer's equipment or with earlier versions of this manufacturer's equipment; to sort or merge data files; to initialise new storage media such as tapes or disks.

UV-erasable See *EPROM*.

318

V

V series recommendations A series of recommendations relating to data communication over the telephone network, including the transmission speeds allowed and the operational modes used. They are endorsed by the world's telecommunications authorities through their standards-making body, CCITT.

Compare *X series recommendations*.

More detail *V.21*; *V.22*; *V.23*; *V.24*; *V.29*; *V.32*.

V.21 A CCITT recommendation specifying a mode of operation for modems transmitting data across a telephone network. V.21 modems provide full duplex asynchronous transmission at up to 300 bit/s, and are generally used for low speed transmission from keyboard or visual display terminals, and from personal computers.

V.22 A CCITT recommendation specifying a mode of operation for modems. V.22 modems provide full duplex transmission at 1200/1200 bit/s, with an optional lower speed of 600 bit/s. They will accept either synchronous or asynchronous data, although only asynchronous data can be transmitted over the telephone network. They are used both for interactive (question and answer) working and for file transfer. An enhanced version, known as V.22 bis, increases transmission speed to 2400/2400 bit/s.

V.23 A CCITT recommendation specifying a mode of operation for modems transmitting data across a telephone network. V.23 modems are commonly used for applications such as information retrieval from a remote computer or database, where most of the traffic flows in one direction. Requests for information are sent via a 75 bit/s channel, which matches normal typing speed, and the data is returned at a higher speed of 1200 bit/s. (This is normally represented as 1200/75 bit/s.)

V.24 A CCITT recommendation defining the interface – the pin connections for different signals, etc. – between a modem and a computer system. Functionally identical to the RS-232 standard.

V.29 A CCITT recommendation specifying a mode of operation for modems transmitting data across point-to-point telephone lines. V.29 modems transmit at 9600 bit/s simplex (i.e. one way only).

V.32 A CCITT recommendation specifying a mode of operation for modems transmitting data across a telephone network. V.29 modems transmit at 9600 bit/s duplex (i.e. 9600/9600 bit/s).

vaccine A piece of software that prevents a virus from gaining entry to a computer system and/or removes the virus code from 'infected' disks.

VADS Abbreviation of *value-added data services*.

validate Check that values are within an expected range and hence probably not erroneous.

value-added data services (VADS) A category of services based on a value-added network which enable organisations to exchange data with one another. See, for example, *electronic data interchange*.

value-added network (VAN) A data communications network providing a service of value over and above transport of information alone. The added value may consist in converting from one transmission speed or character code to another, so that different types of device can exchange information, or in providing network management – connecting users, collecting traffic statistics, etc. Services provided via value-added networks are regulated, to prevent infringement of the monopoly in basic (that is, non-value-added) telecommunications services. Two classes of value-added network have been defined for regulatory purposes – value-added data services (VADS) and managed data network services (MDNS).

value added reseller (VAR) A business organisation that buys the computer equipment from a manufacturer, then adds value in the form of applications packages and support skills, and sometimes additional hardware components, before selling on to a customer. Value added resellers are usually appointed by major manufacturers to sell to specialist markets where they themselves lack the resources or the expertise to market effectively.

valve See *thermionic valve*.

VAN Abbreviation of *value-added network*.

vapourware A term used ironically to describe products that are announced well before they are available for release. It is not unusual for such products to be delivered later and in a less exciting form than when first announced, or even for them to fail to appear at all.

VAR Abbreviation of *value added reseller*.

variable A data field whose value changes during execution of a program. Source programs refer to variables by means of identifiers that are defined before or as a variable is first referenced.

Compare *constant*.

variable-length record A record whose length may vary throughout a file. The length of each record may either be defined within the record, such as by a word or character count at the beginning. Alternatively, each record may be terminated by a special character. Records in ASCII text files, for example, are terminated by a carriage return character.

Compare *fixed-length record*.

variant field One of a choice of several alternative fields which may appear in a data record (normally as the last field of the record).

VDM Abbreviation of *Vienna development method*.

VDT Abbreviation of *visual display terminal*.

VDU Abbreviation of *visual display unit*.

vector font See *outline font*.

vector graphics See *object graphics*.

vectorisation A process, applied by software to scanned information such as a map, that interprets the information and creates a digital representation of its structure. Vectorisation is not as accurate as manual digitising, in other words keyboard entry of coordinates and so on, but it saves a great deal of time and labour. It is sometimes used in conjunction with interactive editing, with improved accuracy.

version A means of identifying the release of a piece of software. In order to keep control of different versions of an applications package, version numbers are given to each successive release. Usually a decimal number is used to identify each version – a change in the integer means a major revision (which may not be fully compatible with earlier versions, requiring programs or files to be converted), and a change in the decimal means a minor revision, such as to correct bugs.

vertical blanking interval The delay between the sending of successive pictures in a television transmission, originally included to enable the electron scanning beam to return to the top of the screen. This interval in transmission is used to transmit the frames

321

included in broadcast teletext transmissions, such as the BBC's Ceefax and ITv's Oracle.

vertical redundancy check See *parity bit*.

vertical tabulation Movement of the cursor on a display, or the print mechanism of a printer, vertically downwards, either by a specified number of lines or to preset positions.

very large scale integration (VLSI) Production of integrated circuits each containing tens of thousands of components. The terminology used to describe the scale of integration has already been left behind by progress. At present, four million bit memory (dynamic RAM) chips, containing about 16 million components, are already in use. Effort is now being directed at wafer scale integration – interconnecting a number of chips on a single wafer – as well as at increasing densities on individual chips.

VESA Abbreviation of *video electronics standards association*.

VGA Abbreviation of *video graphics array*.

VGA 8 See *video graphics array*.

video (1) Information in the form of moving pictures. Full motion video, as in a normal television broadcast, requires a bandwidth of several million bits for transmission, but freeze frame video, in which the picture is updated once every few seconds, can be transmitted over a normal telephone line.
(2) Pertaining to a display screen as in, for example, *video display board*.

video conferencing Teleconferencing using video links between the sites participating in the conference. The video pictures at each site may change continuously as in a normal television picture or every few seconds or every few minutes, known as freeze frame, slow scan or stop action video. Freeze frame video is cheaper to install and transmit than full motion video, and can operate over ordinary telephone lines. It is used where it is not necessary to have a continuous record of activity, such as to show slides or pictures accompanying a verbal explanation or for security monitoring.

video display board An expansion card for a personal computer that drives a display screen. As well as being designed for different sizes and designs of display, they are distinguished in particular by the number of greys or colours they can produce on the screen. Boards for monochrome displays, of course, support only two, but boards for colour displays typically support 256, about 65,000, or 16.7 million different colours. These are commonly referred to as 8-bit, 16-bit and 24-bit boards respectively, this being the number of bits needed to hold colour information for each pixel (i. e. point) on the screen.

322

video electronics standards association (VESA) An association of manufacturers of display screens. The association has originated the VL standard for local bus video.

video graphics array (VGA) A standard for the display of information on the screen of a personal computer, introduced by IBM for the larger machines in its PS/2 range. Text characters are constructed from a 9 x 16 matrix of dots. There are two graphics modes: medium (256 colours or 64 grey shades, with a resolution of 320 x 200 pixels) and high (16 colours, 640 x 480). A variant of VGA known as *VGA 8* allows 256 colours on a 360 x 480 screen. VGA also supports programs written for the standards adopted on the earlier PC range.

video memory Memory within a personal computer that records the current contents of the display screen. The size of this memory varies depending on whether the screen displays only black and white, a number of grey shades or colour, since this determines how many bits are needed to record the contents of each pixel on the screen. So that it can easily be upgraded when required, video memory is often separate from the computer's normal memory (RAM), which holds programs and data.

video overlay adaptor A type of expansion card for personal computers that allows full-motion video to be displayed on the computer's screen. It is so called because it allows text and graphics generated by the computer to be overlaid on the video signal.

video scanner A piece of equipment used in desktop publishing that enables a signal from a video device, such as a video camera or a VCR, to be input to a computer. It normally works in conjunction with software in the computer that displays the moving images on the screen, and allows the user to select and freeze particular images. These images can then be manipulated in whatever way is required, such as by altering contrast or by centralising the image, and can then be used for printing. Also known as *image grabber*.

videodisk A disk storage medium that holds video information in analog form. The disk is made of transparent plastic up to 12 inches in diameter. It holds the video signal in a spiral track on the inner surface of the disk, as a series of pits of varying length. These pits are read by laser and represent the frequency-modulated carrier wave used to hold the sound and vision of a video signal. The disk can be used to hold up to 55,500 still images, and the videodisk player can be coupled to a computer that makes use of the pictures on the disk – see *interactive video*.

videotex A generic term for two-way information retrieval services designed for adapted television sets. These sets contain a special decoder that displays on the television screen pages (i.e. screenfuls) of information, retrieved from a remote computer system. A conventional visual display terminal or a personal computer, running special software, can also be used to access a videotex service. Videotex, originally called *viewdata*, was

323

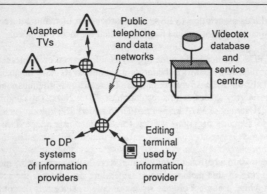

invented by a British Post Office engineer called Sam Fedida in 1970.

Users dial in to the computer system on a normal telephone line and select pages by pressing keys on a keypad, similar to a normal television remote control, which are then displayed on the television screen. Normally, the user works down through index pages to the required information by selecting items from menus that they display. Keys can also be pressed to make selections from menus of products or services, such as to make a booking or to place an order. Videotex is easy to use, because it has few operating instructions and requires no special training, and is designed for casual use by the public, both at home and in a business context.

A number of countries have established public videotex services, such as British Telecom's *Prestel* in the UK.

Companies can buy space in the Prestel database and enter pages containing whatever information they wish to make available to people with videotex sets. Travel and transport companies, for example, provide details of booking status that can be accessed by travel agents. Such companies are known as *information providers*, while the organisation operating the videotex system, in this case British Telecom, is known as the *service provider*. Companies can also connect their own computer systems into Prestel, so that pages can be supplied directly from there, rather than being entered separately. Some companies have also established private videotex services, for example to distribute corporate information internally or to give customers details of their current product range.

videotex decoder See *decoder*.

Vienna development method (VDM) A formal method for specifying and designing application systems. It was originated at the Vienna laboratories of IBM as a mechanism for defining the structure and meaning of programming languages, but this original use has been overtaken by its value as a method of developing applications software. To use the method, an application is first described in terms of the objects that are involved (such as products) and the operations that are applied (such as removing from stock). The objects are then defined using a branch of mathematics called set theory that deals with collections of objects, and the effect of operations is clarified by specifying pre- and post-conditions, in other words what must be true before and after an operation is carried out. This process is repeated for the design of programs, and at each stage the design can be validated mathematically to verify that it is an accurate reflection of the previous stage.

viewdata The term originally chosen by British Telecom for *videotex*. Now superseded.

virtual circuit A connection between two devices across a packet-switching network. It is known as a virtual circuit because the two devices using the connection share the transmission path between them with other devices using the network, rather than having exclusive use of it as they do of a telephone circuit.

virtual machine An environment for programs running within a computer, created by the operating system, which makes it appear to each of them that it has control of an independent machine, providing whatever resources it requires.

virtual memory A technique used in multiprogramming computers to make best use of memory. Instead of being held in memory all the time they are running, programs, together with their internal data, are divided up into smaller areas called pages. Whenever additional memory is needed, because a new program is loaded or an existing program asks for more memory to store data, the operating system decides which pages are least likely to be needed soon, using an algorithm based on most recent use, frequency of use, and priority of the program concerned. These pages are written to disk storage, and the memory they occupied is made available to meet the demand for additional memory. Subsequently, when a program which has pages on disk attempts to use these, they are brought back into memory, displacing other pages in the same way that they themselves were displaced earlier. Virtual memory is advantageous because there are limits on the amount of memory which processors can address, and also because the cost of disk space is much cheaper than that of memory.

virtual screen A mechanism used on some personal computers to enable the user to switch rapidly between different applications. Several applications can occupy the machine, each one operating as if it has full control of the screen and keyboard. At any one time, only one application is actually using the screen, and the operating software keeps in memory a copy of the latest screen contents for the other applications. When the user wants to switch applications, the screen belonging to the application currently running is 'rolled out' and the stored screen contents of the chosen application is 'rolled in', ready to continue processing.

virtual storage See *virtual memory*.

325

virtual system An environment for programs running within a computer, created by the operating system, which enables them to operate exactly as they would if running under the control of another operating system. For new ranges of computer, suppliers develop operating systems capable of providing virtual systems for earlier ranges, at the same time as they run programs written specifically for the new range. This enables users to upgrade their equipment to the new range and take advantage of its extra features, but without being forced to re-develop all their applications programs first.

virtual terminal (VT) A concept designed to make applications programs independent of the particular type of terminal they are dealing with. The idea is to define a kind of universal terminal – the virtual terminal – and to represent this definition in standard interconnection protocols. Applications programs would all be written as if they were exchanging messages with this universal terminal. Real terminals would either be designed so that the interface to them looked the same as that of the universal terminal, or conversion software would be written to map between the protocols used by the virtual and actual terminals. A protocol, called the virtual terminal (or VT) protocol, has been approved by ISO within the open systems interconnection (OSI) family (coded ISO 9041).

virus A piece of software code that hides itself, uninvited, within a computer system. What each virus does depends on what the author intends. Generally, they run automatically when certain operations are carried out or at a given time, sometimes causing catastrophic damage, such as by destroying the contents of disks, sometimes producing harmless (but annoying) messages or displays on the screen. A virus is usually introduced on a disk or via a telephone line, and then replicates itself from disk to disk or from computer to computer. For each virus, a 'vaccine' – in other words another piece of software that prevents the virus from doing its damage – can generally be devised as soon as it is clear how a particular virus works.

visible bit image See *bit image*.

visible records computer (VRC) A small computer system, now obsolete, that records information both magnetically and visually on special cards. These cards can both be used as a manual reference file, and also be run through the computer system to be updated automatically.

visual display See *display screen*.

visual display terminal (VDT) A terminal device equipped with a keyboard and a visual display screen, and containing the logic necessary to connect it to a computer system, so that information can be exchanged between them. Often abbreviated to VDT, and sometimes also called a *visual display unit* or VDU.

326

visual display unit (VDU) Synonym for *visual display terminal*.

VLSI Abbreviation of *very large scale integration*.

voice Used to refer to the representation of human speech as electrical signals, as in 'voice channel' or 'voice traffic'.

voice channel A transmission channel designed to carry voice signals, in other words a telephone call. Data can be transmitted over a voice channel using modems, at speeds up to 9,600 bits per second.
　See also *analog*.
　Compare *digital*.

voice data entry Entry of data to a computer system by verbal commands, rather than via a keyboard or in physical form such as punched cards. Voice data entry is used where the user needs his or her hands for some other purpose, such as to operate a machine tool. Most voice data entry systems in use today can only understand a limited number of separate words. Some, described as 'speaker dependent', have to be 'trained' by the users to respond to their own voice patterns. 'Speaker independent' systems can cope with a range of different voices, but have a more limited vocabulary and a lower tolerance of background noise and variations in voice quality.

voice-grade line See *voice channel*.

voice input See *voice data entry*.

voice line See *voice channel*.

voice messaging A service that enables people to exchange spoken messages non-synchronously (in other words the two parties to an exchange need not be at their terminals at the same time). Voice messaging services are provided by special computer-based devices attached to public or private telephone exchanges. Subscribers dial in to the service, then use a touchtone phone or a separate keypad to enter commands to play back messages left in their own electronic mailbox and to leave messages in the electronic mailboxes of other subscribers.

voice recognition The ability of a computer system to understand and act on spoken input. Voice recognition faces two major problems:
　(1) how to recognise continuous speech, as it is normally spoken, as opposed to individual, discrete words or phrases;
　(2) how to handle the variations in tone of voice, accent, intonation, etc. from one speaker to another.
　At present, voice recognition devices can only handle a limited vocabulary – 500

words is considered large – and must be fed a series of discrete words, each one punctuated by a pause. They also need to be trained – hearing the user speak each of the words in the vocabulary several times so that they can build up a sound picture of the word. Recognition of continuous speech, such as would be necessary to build a voice-driven word processor, for example, is believed to be years away.

voice response Verbal response by a computer system. The system 'synthesises' its response by assembling digitally coded information into something resembling the human voice, hence the alternative term 'speech synthesis'. Voice response is used mainly in industrial applications, but also to respond to transactions entered by telephone, such as in home banking or mail ordering, and in toys such as Texas Instruments' Speak & Spell device.

See also *audiotex*; *interactive voice response*.

voice synthesis See *voice response*.

voice traffic Traffic on a communications network consisting of human speech, or in other words telephone conversations. The main characteristics of voice traffic, and which distinguish it from most other types of communications traffic, are that it is an analog signal (although it can be converted to digital form for transmission) in a frequency band from about 300 to 3,400 Hz, and requires a continuous channel for the duration of a call. This is in contrast with most non-voice traffic (data, text, image, but not full motion video), which is digital and sent in short bursts.

voice/data integration The handling of voice and data (or non-voice) traffic on a single transmission line, with a single switching system (such as a digital telephone exchange) or with a single terminal (such as a visual display terminal equipped with a telephone handset). Voice/data integration became a buzzword in the second half of the 1970s, when electro-mechanical telephone exchanges started to be displaced by electronic exchanges controlled by digital computers.

voicegram A message consisting of recorded (usually in digital form) speech, sent like a telegram from one person to another over a communications network.

See also *voice messaging*.

voicemail See *voice messaging*.

volatile memory A memory device that loses its contents when the power is switched off. The widely-used semiconductor memory devices of today (RAM) are volatile. Other semiconductor memory technologies, such as read-only memory (ROM) do retain their contents, but this can only be fixed during manufacture or by a special process, while the contents of RAM does need to vary in the normal course of processing.

volatility A measure of rate of change, such as the degree to which records in a data file are added, deleted or changed.

volume A physical unit of storage on a mass storage device, such as a disk pack or a tape reel. It may hold a number of complete files (the logical unit of storage) and/or parts of files.

volume-sensitive tariff A tariff that varies depending on the volume of information transmitted, in other words the number of data bits. Volume-sensitive tariffs were first introduced for packet-switching networks, where the carrying capacity of the network relates heavily to the volume of traffic it carries rather than, as is the case with telephone networks, the number and duration of the calls.

von Neumann architecture The architecture that is the basis of most digital computers in use today. It was originated by the Hungarian-American John von Neumann near the end of the Second World War. Its key feature is a program stored in the computer's memory, with program instructions read and executed one after another. It can only access one storage location in each memory cycle, which means that its performance is limited by the speed of its memory. Programming languages that have grown with it do not allow bulk changes to the memory of the computer. For applications that process very large quantities of data, such as for weather forecasting or pattern recognition, this turns the memory into a bottleneck. Contrast with *parallel processing*.

VRAM See *video memory*.

VRC Abbreviation of *visible records computer*.

VT Abbreviation of *virtual terminal*.

VT-100 protocol An asynchronous protocol for visual display terminals. VT-100 protocol was first defined by Digital Equipment (DEC) for its own terminal, coded the VT-100. The protocol is now widely used for communication between computer systems and visual display terminals.

W

wafer A single, thin slice of silicon which is the unit of production for integrated circuits, otherwise known as silicon chips. Chips are fabricated in batches of hundreds at a time on each wafer, after which the wafer is sawn up to create the individual chips. These are then tested. Faulty chips are discarded and the remainder are packaged in plastic and supplied with the pins which enable them to be mounted on printed circuit boards. The yield from a wafer, in other words the proportion of chips that are useable, is an important determinant of chip cost.

See also *wafer scale integration*.

wafer scale integration Manufacture of some or all of the integrated circuits (chips) required to make up a computer system on a single wafer of silicon. Most of the recent improvements in the price/performance of chips have been achieved by reducing the size of the individual circuit components they contain, so that more of them can be included on a single chip. At present levels of integration, however, this still means that even systems as small as personal computers need a number of separate chips, while a large-scale system has hundreds. This is a disadvantage in performance terms, because the connections that can be made between the chips, via the printed circuit board on which they are mounted, are more limited in number and slower than the connections that can be made on the chip itself or on the wafer on which it is fabricated. For this reason, attempts have been made to manufacture a number of inter-connected chips on a single wafer, rather than, as is normally done, making all the chips on a wafer identical and splitting them up following manufacture. This introduces a new problem of yield. Failures of chips on a wafer always occur because the silicon has imperfections in it. When the wafer is due to be cut up into individual chips this does not matter. The faulty chips are discarded, and the overall yield is enough to make the process economic. When a number of chips are linked together on a wafer, however, the yield becomes more critical. The more chips included in a single component, the greater the chance that it will be flawed. One way to overcome the problem is to make the component fault tolerant. This is done by including redundant chips and circuits within the component and building

330

logic into it so that faults are bypassed automatically. So far, attempts to achieve wafer scale integration have not proved economic compared with increasing integration at the chip level.

wait state The state of a processor that has no useful work to do. A processor goes into wait state, in other words just cycles doing nothing, when it is waiting for another component to complete its task before it can continue. For example, if it takes four processor cycles to retrieve a word of data from memory, the processor has to wait three cycles after initiating the retrieval before it has any data to work on.

walkthrough See *structured walkthrough*.

WAN Abbreviation of *wide area network*.

warm boot See *warm restart*.

warm restart A restart that can be made without going through an elaborate recovery procedure.
   Compare *cold restart*.

warm spot An area on a display screen where colour or brightness levels vary compared with most of the rest.

watchdog timer A device external to a computer system that interrogates it at regular intervals to verify that it is still operating. Watchdog timers are used for systems required to be highly reliable. On detecting that a computer is no longer operating, it may send a signal to a standby computer that immediately takes over from the failed one.

waveform digitisation A technique used to produce synthetic speech. It is a complex approach and very expensive in computer resources, but can reflect many of the idiosyncratic touches of a speaker's voice and, as a result, gives very lifelike results. Using actual recordings of a human voice, many unique waveforms are captured each second, converted from analog to digital, then stored digitally in the speech synthesis unit. To produce synthetic speech, a number of these unique waveforms are reproduced through a digital-to-analog converter, then amplified.

wavefront array See *systolic array*.

WHILE statement A control structure used in some high-level languages to construct a loop within a program. The statement takes the general form 'WHILE condition DO action', indicating that the statements that constitute 'action' are to be repeated while 'condition' holds true.

wicked problem A problem which is poorly formulated in the minds of those trying to solve it – see *issue-based information system*.

wide area network (WAN) A network, or part of a network, that interconnects sites and/or buildings, often over large distances. Contrast with *local area network*, which interconnects devices within a building or on a single site over limited distances.

wideband Used of transmission lines to mean circuits capable of speeds of 48, 64 or 72 kilobits per second.

widget A small program associated with an icon or an object in a graphical user interface. Widgets are included in toolkits intended for software developers, and make it easier for them to develop applications programs using the graphical user interface.

widow A single line at the end of a paragraph that starts a new page or column.
Compare *orphan*.

wildcard character A special character that may be included in a field to be used for a search operation and means that any character in that position may be accepted. For example, if '?' is used as a wildcard character, 'c?t' means any three-character field beginning with 'c' and ending with 't'.

WIMPs Acronym for *windows, icons, mouse, pull-down menus* (or alternatively *windows, icons, menus, pointing*) – the four key elements of the human/computer interface first developed by Xerox Corporation and subsequently popularised by Apple Computer as embodied in the Macintosh personal computer. Such an interface is designed to enable people to communicate with the machine in an intuitive way, rather than have to learn a set of abstract operating rules. It also gives all applications running under it a consistent 'look and feel', which makes it easier for users to learn new applications. Also known as *graphical user interface*.

Winchester disk A hard disk widely used with personal computers. It is physically small, so that it can be fitted inside the processor unit of a personal computer or in a box that sits under the screen unit, and has flying heads that float some 10 to 20 microns above the recording surface. It has a storage capacity ranging from ten to several hundred megabytes. The drive is totally sealed, which makes the disk highly reliable but also non-removable. It was invented at IBM's San José laboratories in the 1970s. 'Winchester' was IBM's internal code name for the development project.

window A rectangular area on a visual display screen, used to hold the information relating to a particular aspect of an application currently in operation. Windows overlap one another on the screen as if they were sheets of paper overlapping one another on a desk. The topmost window represents the application with which the user is currently working. A particular window can be brought to the top of the 'pile' by selecting menu options or by moving the cursor to a chosen window and clicking the mouse button.

See also *graphical user interface*.

Compare *split screen*.

window size The number of packets that can be sent through a packet-switching network at a time, before the calling terminal must wait for an acknowledgement from the called terminal.

Windows A graphical user interface for IBM-compatible personal computers, developed by Microsoft. The Windows software sits between the operating system (MS-DOS/PC-DOS) and the applications, providing mechanisms for windowing, dialogue management and so on that the latter can call on as required.

wired building See *intelligent building*.

wired society A society organised around the use of telecommunications networks for communication and for business transactions.

wireframe model A geometric representation of an object by lines and arcs that does not describe the object's physical attributes.

Compare *solid model*.

WKS The proprietary format defined by Lotus for files representing spreadsheets and incorporated into the company's highly successful package, 1-2-3. A number of other spreadsheet packages are able to import files in this format.

word The unit in which information is normally stored into and retrieved from the memory of a computer, and thus handled by the processor as a single entity. Most computers have a fixed word length which is the same as the length of program instructions and a multiple of the length of character codes stored internally – usually 16, 24 or 32 bits.

word length The size of a word as defined for a particular computer, measured by the number of bits it contains.

word processing (WP) An application that enables people to enter, edit, format, manage and print documents consisting of text and, sometimes, simple graphics and tables. Also the act of using such an application. Originally, word processing was carried out on special-purpose devices, sometimes with an A4-sized screen rather than a standard (24 lines by 80 characters) data terminal screen. But increasingly standard personal computers are used, with a range of applications packages. At one end of that range are easy-to-use and relatively limited packages designed for casual typists such as managers; at the other end, very powerful packages that can be used, for example, to take the text of a book all the way from initial draft through to camera-ready copy – here word processing overlaps into desktop publishing. The key to word processing is that documents are held in computer memory and/or on disk storage while they are being prepared, with the part of the document that is currently being processed displayed on a screen, and they are only printed out when the user decides. This means that changes can more easily be made than if, as happens with a typewriter, a hard copy version is produced at the same time as text is entered.

word processor A small computer system designed specifically for word processing. The equipment consists of the elements now familiar in standard personal computers – keyboard, screen, processor (dedicated to a single keyboard/screen or shared between a number of them), disk storage (both floppy and hard) and printer. Some word processors additionally have special features designed for typists, such as A4-sized screens and function keys for cursor control and formatting.

work area An area of memory used by a program to store information temporarily in the course of processing.

work file A data file used for temporary storage of information, such as to hold interim results produced in the course of a program compilation or a sort. Normally, work files are allocated by the operating system when a program requests, and are released automatically as soon as the program terminates.

work group A group of individuals that work together and therefore share and exchange information regularly. May be the same as a department, or alternatively a group within a department. Many work groups, however, have members belonging to other departments and located on other sites.
    See also *departmental computing*.

work group computing See *departmental computing*.

work-related upper limb disorders (WRULD) A term used by ergonomists and medical staff to refer to physical disorders resulting from prolonged intensive work in awkward or constrained postures, such as by using a computer keyboard at a badly designed workplace.
    See also *repetitive strain injury*.

work tape See *work file*.

workflow management See *document routing*.

workstation In general, a powerful (i.e. more than a data terminal) device that can be used to access computing services.

The term was originally used to describe powerful desktop computers equipped with a high-resolution display and designed for technical applications such as computer-aided design, sometimes known as engineering or graphics workstations. Groups of these workstations are normally linked to a shared computer which holds common information, such as details of the parts to be incorporated into designs. The term is increasingly used more broadly to mean any desktop computer system with a high-resolution display screen and able to integrate text and graphics (see *bit-mapped display*), and which can both process and store information locally and exchange information with other systems, as in 'office workstation' and 'management workstation'. The original engineering workstations, such as those introduced by Apollo and Sun in the early 1980s, had substantially more memory and processing power than personal computers, and also were substantially more expensive. Since then, however, both the performance gap and the price differential between these workstations and the standard personal computers used widely in offices has narrowed considerably.

WORM Abbreviation of *Write Once Read Many times*, referring to optical disks on which information cannot be erased or altered once it has been recorded on the disk. Bits are recorded on the disk by lasers, which burn pits or bubbles into a laser-sensitive area on the surface of the disk. One 12-inch WORM disk can hold several gigabytes of information, which equates to hundreds of thousands of pages of text produced on a word processor, or tens of thousands of A4 page images. They are used to store images of business documents, engineering drawings, signatures and the like; as a replacement for tape storage; and for archival storage of data.

See also *CD-WO*.

Compare *CD-ROM*; *erasable optical storage*.

worm A type of virus – a piece of software code that hides itself, uninvited, within a computer system. Rather than causing visible symptoms like other types of virus (such as alarming messages on the screen), it gradually 'eats up' the memory.

WP Abbreviation of *word processing*.

WPR Abbreviation of *write permit ring*.

wraparound Where text is assembled automatically by sofware into lines of a specified maximum length, as it is typed in at a computer keyboard. Words that take the line over that length – either the width of the screen or the required line length for a printed document – 'wrap around' to appear at the beginning of the next line.

write permit ring (WPR) A plastic ring that can be inserted in a groove in a magnetic tape reel, indicating that the reel can be over-written. The tape unit will intercept and prevent write operations to tape reels where the ring is missing, thus protecting the tape against accidental damage as a result of a program or operator error.

write protect Protect (a disk) against accidental overwriting. Floppy disks have a tab that can be slid across or a notch that can be covered over, to indicate to the computer system that it should not attempt to write to the disk.

writing tablet A device used to capture hand-written text and drawings. Writing tablets are used in electronic blackboard systems and to enter drawings into personal computers.

WRULD Abbreviation of *work-related upper limb disorders*.

WYSIWYG Acronym for *what you see is what you get*, pronounced 'wizziwig'. Used to describe word processing software that displays on the screen an accurate representation of what will subsequently be printed on paper.

X

x-height The height of a line of text, excluding descenders.

x-on/x-off A simple method of flow control used for asynchronous transmission, such as between two personal computers. The receiving device sends an 'x-off' control character to stop the transmission of data, and an 'x-on' character to restart it.

X series recommendations A series of recommendations relating to data communication over public data networks, including the allowed transmission speeds, the interfaces to and between networks, and modes of operation for user devices. They are endorsed by the world's telecommunications authorities through their standards-making body, CCITT.
   Compare *V series recommendations*.
   More detail *X.12*; *X.21*; *X.25*; *X.28*; *X.400*; *X.500*.

X user interface (XUI) A set of tools to be used when developing applications to run under X Windows. It provides standard ways of setting up the various mechanisms used in a graphical user interface, such as windows, scroll bars, menus, and so on.

X Windows A standard network protocol for graphics-oriented applications, which incorporates mechanisms for windowing. It was originally devised to meet MIT's requirement for a common vendor-independent graphical user interface for computer systems running Unix, and has since been endorsed by a number of bodies responsible for promoting 'open' standards. Its key feature is its structure, which separates the application program from the terminal (i.e. the screen, keyboard and mouse) at which it will be displayed and controlled. These two parts can communicate, either within the same machine or across a network, using a set of rules known as the X network protocol. The protocol allows the application, known as the client, to instruct the terminal part, which is known as the server, to exchange information with the user, draw on the screen, and so on. Several applications can communicate with a single server, each controlling one or more windows on the screen.

337

X.12 A standard for the formatting of electronic data interchange (EDI) transactions, defined by ANSI and widely used in North America.

X.21 The CCITT standard specifying the interface between a user's device (in tele-communications terminology, the DTE or data terminal equipment) and the modem (the DCE or data communications equipment), for operation over a circuit-switched data network.

X.25 The CCITT standard specifying the interface between a user's device and the modem, for packet mode operation over a packet-switching network. X.25 provides the mechanism for exchanging packets over the network link, including error detection and correction, and defines procedures for setting up and clearing calls.
   More detail *bit stuffing*; *bit-synchronous*; *flag*; *frame*; *packet*; *virtual circuit*.

X.28 The CCITT standard specifying the mode of access to a packet switching network for character mode (i.e. non-packet) devices, using dial-up connections over the tele-phone network. X.28 covers the procedures used to establish an information path between the user's device and the PAD (packet assembler disassembler) in the packet switching exchange; to initialise the service; and to exchange data. There are two associ-ated standards which specify how character mode devices are to be handled: X.3 details how the PAD is to assemble and disassemble packets, and set up and clear down calls; X.29 explains how packet mode terminals (for example, a computer system hosting an information retrieval service) should communicate with the PAD.

X.29 See *X.28*.

X.3 See *X.28*.

X.32 The CCITT standard specifying a synchronous access protocol to packet-switching networks. It offers the convenience of dial-up, over the telephone network or ISDN, together with the performance and reliability of packet transmission.

X.400 (MHS) The CCITT definition of a message handling system, intended for public or private electronic mail services (also known as the *MHS* or Red Book standard). It defines how messages should be transferred across a network or, as will often occur with international messages, across a series of interconnected networks. It does not define how messages should be created or displayed, these tasks being regarded as the local respon-sibility of the receiving and sending terminals.

X.435 A standard for the format of electronic messages, intended particularly for elec-tronic data interchange (EDI). It is a subset of the X.400 standard for message formats.

X.500 The CCITT definition of the directory serviced to be used with X.400 electronic mail services.

X/Open A consortium of European computer manufacturers formed to promote the adoption of 'open' standards. They have defined a Common Applications Environment – a set of standards for the software used to develop and interconnect computer applications. These standards are intended to enable computer users to choose freely from the hardware and software on offer, regardless of supplier, and to create a single, uniform market for independent software suppliers.

See also *open systems interconnection*; *POSIX*.

X3J3 The ANSI committee responsible for defining standards for the FORTRAN programming language.

XGA Abbreviation of *extended graphics array*.

Xmodem A file transfer protocol designed for use over public telephone networks and supported by many bulletin boards and online information services. The original version used checksum error detection to detect transmission errors, initiating repeat transmissions automatically. A later version uses a more effective cyclic redundancy check (CRC). Improved versions called Ymodem and Zmodem have since been developed.

XUI Abbreviation of *X user interface*.

Y

---

year-to-date (YTD) Used to describe rolling totals accumulated weekly or monthly, such as sales or production figures.

yield The proportion of good integrated circuits produced from a wafer.
   See also *wafer scale integration*.

Ymodem An improved version of the Xmodem file transfer protocol. It can send information in larger chunks (up to 1024 characters per block) and can transfer multiple files in a single operation. A variant called *Ymodem-G* is designed for high performance using error-detecting modems, transferring an entire file as a single block.

Yourdon method A so-called 'structured design' method for designing applications systems and programs, originated by the man whose name it bears.

YTD Abbreviation of *year-to-date*.

Z

Z A formal method for specifying application systems, originally devised by the French computer scientist Jean Abrial and considerably expanded by the Programming Research Group at Oxford University. It uses a semi-graphical notation consisting of mathematical logic enclosed in boxes called schemas. Each schema describes some stored data and the effect of operations on that data.

Zmodem A sophisticated file transfer protocol, developed from the long-established Xmodem protocol. It uses a 32-bit cyclic redundancy check (CRC) to detect errors, and a sliding window protocol to take advantage of high-speed modems. It also supports checkpointing, so that an interrupted file transfer can be resumed from the point of breakdown.

zone One local area network in a series of such networks interconnected by *bridges*.

Appendix 1: ASCII code chart

To work out the decimal value of the code for any item, multiply *colum'n* by 16 and add **row**. Conversely, to find out the character for a given decimal value, divide it by 16 — the whole number result indicates the column and the remainder the row.

Row	Column							
	0	*1*	*2*	*3*	*4*	*5*	*6*	*7*
0	NUL	DLE	space	0	@	P	`	p
1	SOH	DC1	!	1	A	Q	a	q
2	STX	DC2	"	2	B	R	b	r
3	ETX	DC3	£ or #	3	C	S	c	s
4	EOT	DC4	$	4	D	T	d	t
5	ENQ	TC8	%	5	E	U	e	u
6	ACK	SYN	&	6	F	V	f	v
7	BEL	ETB	'	7	G	W	g	w
8	BS	CAN	(	8	H	X	h	x
9	HT	EM	)	9	I	Y	i	y
10	LF	SUB	*	:	J	Z	j	z
11	VT	ESC	+	;	K	[	k	{
12	FF	FS	,	<	L	\	l		
13	CR	GS	-	=	M	]	m	}
14	SO	RS	.	>	N	^	n	~
15	SI	US	/	?	O	_	o	DEL

Notes

BS = backspace

HT = horizontal tabulation

LF = line feed

FF = form feed

CR = carriage return

ESC = escape

DEL = delete

STX, ETX, EOT, ENQ, ACK, SYN, ETB, CAN – see entries in dictionary

Others in columns 0 and 1 are various transmission control functions

In some countries, positions towards the foot of columns 5 and 7 are used for diacritical signs and special characters

READ MORE IN PENGUIN

In every corner of the world, on every subject under the sun, Penguin represents quality and variety – the very best in publishing today.

For complete information about books available from Penguin – including Puffins, Penguin Classics and Arkana – and how to order them, write to us at the appropriate address below. Please note that for copyright reasons the selection of books varies from country to country.

In the United Kingdom: Please write to *Dept. EP, Penguin Books Ltd, Bath Road, Harmondsworth, West Drayton, Middlesex UB7 ODA*

In the United States: Please write to *Consumer Sales, Penguin USA, P.O. Box 999, Dept. 17109, Bergenfield, New Jersey 07621-0120.* VISA and MasterCard holders call 1-800-253-6476 to order Penguin titles

In Canada: Please write to *Penguin Books Canada Ltd, 10 Alcorn Avenue, Suite 300, Toronto, Ontario M4V 3B2*

In Australia. Please write to *Penguin Books Australia Ltd, P.O. Box 257, Ringwood, Victoria 3134*

In New Zealand: Please write to *Penguin Books (NZ) Ltd, Private Bag 102902, North Shore Mail Centre, Auckland 10*

In India: Please write to *Penguin Books India Pvt Ltd, 706 Eros Apartments, 56 Nehru Place, New Delhi 110 019*

In the Netherlands: Please write to *Penguin Books Netherlands bv, Postbus 3507, NL-1001 AH Amsterdam*

In Germany: Please write to *Penguin Books Deutschland GmbH, Metzlerstrasse 26, 60594 Frankfurt am Main*

In Spain: Please write to *Penguin Books S. A., Bravo Murillo 19, 1° B, 28015 Madrid*

In Italy: Please write to *Penguin Italia s.r.l., Via Felice Casati 20, 1–20124 Milano*

In France: Please write to *Penguin France S. A., 17 rue Lejeune, F–31000 Toulouse*

In Japan: Please write to *Penguin Books Japan, Ishikiribashi Building, 2–5–4, Suido, Bunkyo-ku, Tokyo 112*

In South Africa: Please write to *Longman Penguin Southern Africa (Pty) Ltd, Private Bag X08, Bertsham 2013*

READ MORE IN PENGUIN

SCIENCE AND MATHEMATICS

The Character of Physical Law   Richard P. Feynman

'Richard Feynman had both genius and highly unconventional style ... His contributions touched almost every corner of the subject, and have had a deep and abiding influence over the way that physicists think' – Paul Davies

A Mathematician Reads the Newspapers   John Allen Paulos

In this book, John Allen Paulos continues his liberating campaign against mathematical illiteracy. 'Mathematics is all around you. And it's a great defence against the sharks, cowboys and liars who want your vote, your money or your life' – Ian Stewart

Bully for Brontosaurus   Stephen Jay Gould

'He fossicks through history, here and there picking up a bone, an imprint, a fossil dropping, and, from these, tries to reconstruct the past afresh in all its messy ambiguity. It's the droppings that provide the freshness: he's as likely to quote from Mark Twain or Joe DiMaggio as from Lamarck or Lavoisier' – *Guardian*

Are We Alone?   Paul Davies

Since ancient times people have been fascinated by the idea of extraterrestrial life; today we are searching systematically for it. Paul Davies's striking new book examines the assumptions that go into this search and draws out the startling implications for science, religion and our world view, should we discover that we are not alone.

The Making of the Atomic Bomb   Richard Rhodes

'Rhodes handles his rich trove of material with the skill of a master novelist ... his portraits of the leading figures are three-dimensional and penetrating ... the sheer momentum of the narrative is breathtaking ... a book to read and to read again' – *Guardian*

READ MORE IN PENGUIN

SCIENCE AND MATHEMATICS

Bright Air, Brilliant Fire Gerald Edelman

'A brilliant and captivating new vision of the mind' – Oliver Sacks. 'Every page of Edelman's huge wok of a book crackles with delicious ideas, mostly from the *nouvelle cuisine* of neuroscience, but spiced with a good deal of intellectual history, with side dishes on everything from schizophrenia to embryology' – *The Times*

Games of Life Karl Sigmund
Explorations in Ecology, Evolution and Behaviour

'A beautifully written and, considering its relative brevity, amazingly comprehensive survey of past and current thinking in "mathematical" evolution ... Just as games are supposed to be fun, so too is *Games of Life*' – *The Times Higher Education Supplement*

The Artful Universe John D. Barrow

In this original and thought-provoking investigation John D. Barrow illustrates some unexpected links between art and science. 'Full of good things ... In what is probably the most novel part of the book, Barrow analyses music from a mathematical perspective ... an excellent writer' – *New Scientist*

The Doctrine of DNA R. C. Lewontin

'He is the most brilliant scientist I know and his work embodies, as this book displays so well, the very best in genetics, combined with a powerful political and moral vision of how science, properly interpreted and used to empower all the people, might truly help us to be free' – Stephen Jay Gould

Artificial Life Steven Levy

'Can an engineered creation be alive? This centuries-old question is the starting point for Steven Levy's lucid book ... *Artificial Life* is not only exhilarating reading but an all-too-rare case of a scientific popularization that breaks important new ground' – *The New York Times Book Review*

READ MORE IN PENGUIN

SCIENCE AND MATHEMATICS

About Time   Paul Davies

'With his usual clarity and flair, Davies argues that time in the twentieth century is Einstein's time and sets out on a fascinating discussion of why Einstein's can't be the last word on the subject' – *Independent on Sunday*

Insanely Great   Steven Levy

It was Apple's co-founder Steve Jobs who referred to the Mac as 'insanely great'. He was absolutely right: the machine that revolutionized the world of personal computing was and is great – yet the machinations behind its inception were nothing short of insane. 'A delightful and timely book' – *The New York Times Book Review*

Wonderful Life   Stephen Jay Gould

'He weaves together three extraordinary themes – one palaeontological, one human, one theoretical and historical – as he discusses the discovery of the Burgess Shale, with its amazing, wonderfully preserved fossils – a time-capsule of the early Cambrian seas' – *Mail on Sunday*

The *New Scientist* Guide to Chaos   Edited by Nina Hall

In this collection of incisive reports, acknowledged experts such as Ian Stewart, Robert May and Benoit Mandelbrot draw on the latest research to explain the roots of chaos in modern mathematics and physics.

Innumeracy   John Allen Paulos

'An engaging compilation of anecdotes and observations about those circumstances in which a very simple piece of mathematical insight can save an awful lot of futility' – *The Times Educational Supplement*

Consciousness Explained   Daniel C. Dennett

'Extraordinary ... Dennett outlines an alternative view of consciousness drawn partly from the world of computers and partly from the findings of neuroscience. Our brains, he argues, are more like parallel processors than the serial processors that lie at the heart of most computers in use today ... Supremely engaging and witty' – *Independent*

READ MORE IN PENGUIN

POPULAR SCIENCE

Naturalist Edward O. Wilson

'One of the finest scientific memoirs ever written, by one of the finest scientists writing today' – *Los Angeles Times*. 'There are wonderful accounts of his adventures with snakes, a gigantic ray, butterflies, flies, and, of course, ants ... provides a fascinating insight into a great mind' – *Guardian*

Eight Little Piggies Stephen Jay Gould

'Stephen Jay Gould has a talent for making the scientific, and particularly the evolutionary, interesting and striking ... Time and again in these essays ... he demonstrates the role of the randomness or waste in evolution' – *Sunday Times*. 'His essays are remarkable, and not just for the scientific insights at their centre ... Gould takes his readers on tough-minded rambles across the visible surface of things' – *Guardian*

Darwin's Dangerous Idea Daniel C. Dennett

'A surpassingly brilliant book. Where creative, it lifts the reader to new intellectual heights. Where critical, it is devastating. Dennet shows that intellectuals have been powerfully misled on evolutionary matters and his book will undo much damage' – Richard Dawkins

Relativity for the Layman James A. Coleman

Einstein's Theory of Relativity is one of the greatest achievements of twentieth-century science. In this clear and concise book, James A. Coleman provides an accessible introduction for the non-expert reader.

God and the New Physics Paul Davies

How did the world begin – and how will it end? These questions are not new; what is new, argues Paul Davies, is that we may be on the verge of answering them. 'The author is an excellent writer. He not only explains with fluent simplicity some of the profoundest questions of cosmology, but he is also well read in theology' – *Daily Telegraph*

READ MORE IN PENGUIN

BUSINESS AND ECONOMICS

Trust Francis Fukuyama

'The man who made his name proclaiming the end of history when communism collapsed has now re-entered the lists, arguing that free markets, competition and hard work are *not* the sole precursors for prosperity. There is another key ingredient – trust ... This is the heart of Fukuyama's theory ... it is both important and full of insight' – *Guardian*

I am Right – You are Wrong Edward de Bono

Edward de Bono expects his ideas to outrage conventional thinkers, yet time has been on his side, and the ideas that he first put forward twenty years ago are now accepted mainstream thinking. Here, in this brilliantly argued assault on outmoded thought patterns, he calls for nothing less than a New Renaissance.

Lloyds Bank Small Business Guide Sara Williams

This long-running guide to making a success of your small business deals with real issues in a practical way. 'As comprehensive an introduction to setting up a business as anyone could need' – *Daily Telegraph*

The Road Ahead Bill Gates

Bill Gates – the man who built Microsoft – takes us back to when he dropped out of Harvard to start his own software company and discusses how we stand on the brink of a new technology revolution that will for ever change and enhance the way we buy, work, learn and communicate with each other.

Exploring Management Across the World David J. Hickson

This companion volume to *Management Worldwide* contains selections from seminal writings on centralization, individualism, work relationships, power and risk among managers and countries and cultures all over the globe.

Understanding Organizations Charles B. Handy

Of practical as well as theoretical interest, this book shows how general concepts can help solve specific organizational problems.

READ MORE IN PENGUIN

BUSINESS AND ECONOMICS

In with the Euro, Out with the Pound   Christopher Johnson

The European Union is committed to setting up the Euro as a single currency, yet Britain has held back, with both politicians and public unable to make up their minds. In this timely, convincing analysis, Christopher Johnson asserts that this 'wait and see' policy is damaging and will result in far less favourable entry terms.

Lloyds Bank Tax Guide   Sara Williams and John Willman

An average employee tax bill is over £4,000 a year. But how much time do you spend checking it? Four out of ten never check the bill – and most spend less than an hour. Mistakes happen. This guide can save YOU money. 'An unstuffy read, packed with sound information' – *Observer*

The Penguin Companion to European Union
Timothy Bainbridge with Anthony Teasdale

A balanced, comprehensive picture of the institutions, personalities, arguments and political pressures that have shaped Europe since the end of the Second World War.

Understanding Offices   Joanna Eley and Alexi F. Marmot

Few companies systematically treat space as a scarce resource or make conscious efforts to get the best from their buildings. This book offers guidance on image, safety, comfort, amenities, energy-efficiency, value for money and much more.

Faith and Credit   Susan George and Fabrizio Sabelli

In its fifty years of existence, the World Bank has influenced more lives in the Third World than any other institution, yet remains largely unknown, even enigmatic. This richly illuminating and lively overview examines the policies of the Bank, its internal culture and the interests it serves.

READ MORE IN PENGUIN

REFERENCE

The Penguin Dictionary of Literary Terms and Literary Theory
J. A. Cuddon

'Scholarly, succinct, comprehensive and entertaining, this is an important book, an indispensable work of reference. It draws on the literature of many languages and quotes aptly and freshly from our own' – *The Times Educational Supplement*

The Penguin Dictionary of Symbols
Jean Chevalier and Alain Gheerbrant, translated by John Buchanan-Brown

This book draws together folklore, literary and artistic sources and focuses on the symbolic dimension of every colour, number, sound, gesture, expression or character trait that has benefited from symbolic interpretation.

Roget's Thesaurus of English Words and Phrases
Edited by Betty Kirkpatrick

This new edition of Roget's classic work, now brought up to date for the nineties, will increase anyone's command of the English language. Fully cross-referenced, it includes synonyms of every kind (formal or colloquial, idiomatic and figurative) for almost 900 headings. It is a must for writers and utterly fascinating for any English speaker.

The Penguin Guide to Synonyms and Related Words
S. I. Hayakawa

'More helpful than a thesaurus, more humane than a dictionary, the *Guide to Synonyms and Related Words* maps linguistic boundaries with precision, sensitivity to nuance and, on occasion, dry wit' – *The Times Literary Supplement*

The Penguin Book of Exotic Words   Janet Whitcut

English is the most widely used language today, its unusually rich vocabulary the result of new words from all over the world being freely assimilated into the language. With entries arranged thematically, words of Saxon, Viking, French, Latin, Greek, Hebrew, Arabic and Indian origin are explored in this fascinating book.

READ MORE IN PENGUIN

REFERENCE

Medicines: A Guide for Everybody Peter Parish

Now in its seventh edition and completely revised and updated, this bestselling guide is written in ordinary language for the ordinary reader yet will prove indispensable to anyone involved in health care – nurses, pharmacists, opticians, social workers and doctors.

Media Law Geoffrey Robertson QC and Andrew Nichol

Crisp and authoritative surveys explain the up-to-date position on defamation, obscenity, official secrecy, copyright and confidentiality, contempt of court, the protection of privacy and much more.

The Penguin Careers Guide
Anna Alston and Anne Daniel; Consultant Editor: Ruth Miller

As the concept of a 'job for life' wanes, this guide encourages you to think broadly about occupational areas as well as describing day-to-day work and detailing the latest developments and qualifications such as NVQs. Special features include possibilities for working part-time and job-sharing, returning to work after a break and an assessment of the current position of women.

The Penguin Dictionary of Troublesome Words Bill Bryson

Why should you avoid discussing the *weather conditions*? Can a married woman be celibate? Why is it eccentric to talk about the aroma of a cowshed? A straightforward guide to the pitfalls and hotly disputed issues in standard written English.

The Penguin Dictionary of Musical Performers Arthur Jacobs

In this invaluable companion volume to *The Penguin Dictionary of Music* Arthur Jacobs has brought together the names of over 2,500 performers. Music is written by composers, yet it is the interpreters who bring it to life; in this comprehensive book they are at last given their due.

READ MORE IN PENGUIN

DICTIONARIES

Abbreviations
Ancient History
Archaeology
Architecture
Art and Artists
Astronomy
Biology
Botany
Building
Business
Challenging Words
Chemistry
Civil Engineering
Classical Mythology
Computers
Curious and Interesting Geometry
Curious and Interesting Numbers
Curious and Interesting Words
Design and Designers
Economics
Electronics
English and European History
English Idioms
Foreign Terms and Phrases
French
Geography
Geology
German
Historical Slang
Human Geography
Information Technology

International Finance
Literary Terms and Literary Theory
Mathematics
Modern History 1789–1945
Modern Quotations
Music
Musical Performers
Nineteenth-Century World History
Philosophy
Physical Geography
Physics
Politics
Proverbs
Psychology
Quotations
Religions
Rhyming Dictionary
Russian
Saints
Science
Sociology
Spanish
Surnames
Symbols
Telecommunications
Theatre
Third World Terms
Troublesome Words
Twentieth-Century History
Twentieth-Century Quotations